CONCISE DICTIONARY
OF ENGINEERING

D1600172

CONCISE DICTIONARY
OF ENGINEERING

A GUIDE TO THE LANGUAGE OF ENGINEERING

RON HANIFAN

Del Rio, TN, USA

 Springer

Ron Hanifan
Del Rio, TN, USA

Portions of this Work previously appeared in The Engineering Language, 978-1-606-50206-8 (originally published in 2010 by Momentum Press).

ISBN 978-3-319-07838-0   ISBN 978-3-319-07839-7 (eBook)
DOI 10.1007/978-3-319-07839-7
Springer Cham Heidelberg New York Dordrecht London

Library of Congress Control Number: 2014942313

Printed on acid-free paper

Springer is part of Springer Science+Business Media (www.springer.com)

PREFACE

In this book, I have provided many of the most common terms and words that are used throughout the engineering community along with their proper definition. These are recognized and approved definitions, and they have been extracted from appropriate sources such as military handbooks, standards, and specifications, and commercial specifications, and consolidated into one document.

It is imperative that when a word or term is used all parties understand the proper definition. Ours is a structured field, and any ambiguity has the potential of costing millions of dollars. Often, words or terms and their meanings get intermixed or misapplied. Many of these words/terms are used in engineering documents, verbal communication, contracts, purchase orders, technical orders, technical manuals, and all documentation related to or associated to the engineering field. The words we use form a legal agreement, so when used improperly, repercussions will result.

Notes on drawings and any associated documents are to be written in short, concise statements, using the simplest words and phrases for conveying the intended meaning. The intended meaning will be understood when words/terms are used properly. When an incorrect application of a word/term is used, it can be disastrous.

Ron Hanifan
Del Rio, TN, USA

ACKNOWLEDGMENTS

This book is dedicated to my lovely wife, Catherine. She is the inspiration and driving force behind all that I write. She gave me the encouragement to finish this book and to leave my little mark in this beautiful world. Semper Fi.

CONTENTS

ABOUT THE AUTHOR

Ron Hanifan has over 30 years of experience in the engineering field and has been active in many major defense and commercial programs with some of the top agencies in the United States. He has developed projects at Lockheed Martin, Magnavox, General Dynamics, Teledyne Brown, Litton Data Systems, Motorola, and Borg Warner.

His experience includes positions as Staff Engineer, Electrical and Mechanical Design Analyst, Mechanical Designer, Engineering Procedures Specialist, Customer Contractor Negotiator, and instructor for Technical Data Packages Requirements and Geometric Dimensioning and Tolerancing.

A

Abbe Constant
Ration of refractivity of a material to its dispersion. Also called the Nu or Vee value. A calculated value from known indices of refraction.

Abbreviation
An abbreviation is a shortened form or abridgment of a word, expression or phrase used to conserve space and time. Syntax: The same abbreviation is used for all tenses, the possessive case, participle endings, the singular or plural, and the noun modifying forms of a term. Used only when their meanings are unquestionable clear to the intended reader. When they are not in common use and are to be used frequently in a document or where there is doubt that the reader will understand their meanings, it should be clarified when first used.

Aberration
(1) Generally, any systematic deviation from an ideal path of the image-forming rays passing through an optical system that causes the image to be imperfect. Specific aberrations are spherical aberrations, coma, chromatic aberration, and distortion. (2) The phenomenon caused by rotation of the earth in its orbit that is sufficient to cause the light from a star to appear to shift forward. For those starting at right angles to the direction of the earth's travel, the maximum effect is 20.5 seconds of arc.

Aberration, Chromatic
Image imperfection caused by light of different wavelengths following different paths through an optical system due to dispersion in the optical elements of the system.

Aberration, Chromatic, Lateral
A variation in the size of images for lights of different colors or wavelengths produced by an optical system. It is measured as the radial displacement of the image in the first color, from the image of the same point in the second color. A direction away from the axial image point is taken as positive, and a direction toward the image point is of difference in magnification for light of different colors, and is measured as the angular separation in apparent field between the images of the same point in two colors.

Aberration, Chromatic, Longitudinal
The distance between the foci for light of different colors measured along the optical axis.

Aberration, Spherical
A symmetrical optical defect of lenses and spherical mirrors in which light rays that come from a common axial point, but strike the lens at different distances from the optical axis, do not come to a common focus.

Aberration, Spherochromatic
The variation in spherical aberration for light of different wavelengths or colors. Often shown as a graph of the spherical aberration for several specific wavelengths, usually defined by the Fraunhofer lines of the solar spectrum.

Ablation
The wearing away of surface material due to the action of a fluid moving past it at either high speed or high temperature, or both. Ablation is a method of removing the aerodynamic

heating effects on missile nose cones during re-entry. Using the ablative approach, the nose cone material is subjected to thermal degradation involving pyrolysis, depolymerization, melting, vaporization, and combustion with a resulting continuous removal of material.

Ablative Materials
Special materials on the surface of a spacecraft that can be sacrificed, i.e., carried away or vaporized during re-entry into the atmosphere. Kinetic energy is dissipated, and excessive heating of the main structure of the spacecraft is prevented.

Ablative Plastic
A material that absorbs heat (while part of it is being consumed by heat) through a thermal decomposition process known as pyrolysis, which takes place in the near surface layer exposed to heat.

Abort
A capability that cancels all user entries in a defined transaction sequence.

Abrasion Soldering
A soldering process variation in which the faying surface of the base metal is mechanically abraded during soldering.

Abrasive
Material such as sand, crushed chilled cast iron, crushed steel grit, aluminum oxide, silicon carbide, flint, garnet, or crushed slag used for cleaning or surface roughening.

Absolute Altitude
The height or altitude of an object considering, as a base of reference, the surface terrain over which the object is located.

Absolute Coordinates
The values of the X, Y, or Z coordinates with respect to the origin of the coordinate system.

Absolute Data
Values representing the absolute coordinates of a point or other geometric entity on a display surface. The values may be expressed in linear units of the display or of the engineering drawing.

Absolute Delay
The time interval or phase difference between transmission and reception of a signal.

Absolute Error
The magnitude of the error without regard to an algebraic sign.

Absolute Machine Code
Machine language code in which addresses are specified in terms of actual machine locations.

Absolute Measurement
The measurement in which the comparison is directly with quantities whose units is basic units of the system. *Notes*: A. For example, the measurement of speed by measurements of distanced and time is an absolute measurement, but the measurement of speed by a speedometer is not an absolute measurement. B. The word *absolute* implies nothing about precision or accuracy.

Absolute Vector
A directed line segment whose end point is expressed in "absolute coordinates;" contrast with relative vector.

Absolute Velocity
The highest velocity theoretically attainable.

Absolute Zero
The theoretical temperature of a body cooled to the point that no more heat can be extracted. The zero point of thermodynamic temperature scales. By definition of the International Temperature Scale of 1948, $-273.15°C$. Also a point of temperature equal to $-459.69°F$.

Absorber
A material that causes the irreversible conversion of the energy of an electromagnetic wave into another form of energy as a result of its interactions with matter.

Absorption
A process in which one material (the absorbent) takes in or absorbs anther (the absorbate).

Absorption Loss
The loss of power in a transmission circuit that results from the coupling to a neighboring circuit.

Absorption Modulation
Amplitude modulation of the output of a radio transmitter by means of variable-impedance circuitry that is caused to absorb carrier power in accordance with the modulating wave.

Abstraction
A mechanism for hierarchic, stepwise refinement of detail by which it is possible, at each stage of development, to express only relevant detains and to defer (and, indeed, hide) non-relevant details for later refinement.

Accelerated Aging
Laboratory method of speeding up the deterioration of a product in order to estimate its long-time storage and use characteristics.

Accelerated Life Test
A test in which certain factors, such as voltage, temperature, and so forth, are increased or decreased beyond normal operating values to obtain observable deterioration in a reasonable period of time and thereby afford some measure of the probable life under normal operating conditions, or some measure of the durability of the equipment when exposed to the factors being aggravated.

Accelerating Potential
The potential in electron beam welding that imparts velocity to the electrons, thus giving them energy.

Acceleration Blowout
Inadvertent loss of combustion in a turbojet engine incident to an attempted acceleration and over-rich fuel mixture. Also called *flameout*.

Acceleration Feedback
A sensing system for control of a missile by elimination of body bending effects and maintenance of angles of attack at predetermined values. Accelerometers are used to sense body accelerations, which are fed into the control systems for correction of the motion.

Accelerator
A material that, when mixed with a catalyzed resin, will speed up the curing reaction.

Accelerometer, Guided Missile
An instrument designed to measure one or more acceleration components of a guided missile.

Accelerometer, Integrating
An acceleration measuring device whose output signals are either proportional to vehicle velocity or to distance traveled, and which performs an integrating function to achieve these outputs. When used in a rocket or missile, the integrating accelerometer may be preset to reduce fuel flow when the programmed speed has been achieved.

Acceptability Criteria
A limit or limits placed on the degree of nonconformance permitted in material, expressed in definitive operational terms.

Acceptable Quality Level (AQL)
(1) When a continuous series of lots is considered, the AQL is the quality level that, for the purposes of sampling inspection, is the limit of satisfactory process average. (2) The maximum percentage defective (or the maximum number of defects per hundred units) that, for the purposes of sampling inspections, can be considered satisfactory as a process average.

Acceptable Weld
A weld that meets all the requirements and the acceptance criteria prescribed by the welding specifications.

Acceptance
(1) The act of an authorized representative of the Government by which the Government assumes for itself, or as an agent of another, ownership of existing and identified supplies tendered, or approves specific services rendered, as partial or complete performance of the contract on the part of the contractor. (2) Acknowledgment that a product or data conforms to contract content and quality requirements. Acceptance may occur before, at, or after delivery. Upon acceptance, customer assumes ownership.

Acceptance Angle
In fiber optics, half the vertex angle of that cone within which optical power may be coupled into bound modes of an optical fiber. *Note*: power may be coupled into leaky modes and angles exceeding the acceptance angle.

Acceptance Cone
In fiber optics, that cone within which optical power may be coupled into the bound modes of an optical fiber. The acceptance cone is derived by rotating the acceptance angle (the maximum angle within which light will be coupled into a bound mode) about the fiber axis.

Acceptance Criteria
The quality provisions, including inspection and test requirements, that establish the acceptability of an item. It may range from testing the item in its use environment, to verification of electrical/mechanical characteristics, to a simple visual inspection. Criteria that a set of software must satisfy in conformance with deliver requirements. Software delivered for interim operations with discrepant items is said to be accepted with "liens."

Acceptance Number
The maximum number of defects or defective units in the sample that will permit acceptance of the inspection lot or batch.

Acceptance Testing
Testing to verify acceptance criteria for program certification. This is a self-defining term utilized in software and/or hardware producing contracts. The products pass/fail criteria are predetermined. Failure to meet the standard of the criteria causes rejection of the product.

Acceptance Tests
(1) Those test deemed necessary to determine acceptability of boards and as agreed to by purchaser and vendor. (2) The required formal tests conducted to demonstrate acceptability of an item for delivery. They are intended to demonstrate performance to specification requirements and to act as quality control screens to detect deficiencies of workmanship, material, and quality. (3) A test to verify if the unit under test is operating in accordance with the operational specifications.

Acceptance Trial
A trial carried out by nominated representatives of the eventual military users of the weapon or equipment to determine if the specified performance and characteristics have been met.

Access Category
A class to which a user, program, or process in an automated information system may be assigned, based on the resources or groups of resources each user is authorized to use.

Access Code
The preliminary digits that a user must dial to be connected to a particular outgoing trunk group or line.

Access Control
(1) A technique used to define or restrict the rights of individuals or application programs to obtain data from, or place data onto, a storage device. (2) A service feature or technique used to permit or deny use of the components of a communication system.

Access-Control Mechanisms
Access control mechanisms are mechanisms capable of enforcing rules about who can perform what operations or who can access an object containing certain information.

Access Holes
A series of holes in successive layers each set having a common center or axis. These holes in a multilayer printed board provide access to the surface of the land in one of the layers of the board.

Accessibility
(1) A design feature that affects the ease of admission to an area for the performance of visual and manipulative maintenance. (2) Code possesses the characteristic accessibility to the extent that it facilitates selective use of its parts. (Accessibility is necessary for efficiency, testability, and human engineering.) Accessibility is a reflection of the probability of intentional and accidental breaking into a system. Accessibility is nearly synonymous with security.

Accessible
An item is considered accessible when it can be operated, manipulated, removed, or replaced by the suitably clothed and equipped user with applicable body dimensions conforming to the anthropometric range and database specified by the procuring activity or, if not specified by the procuring activity, with applicable 5th to 95th percentile body dimensions. Applicable body dimensions are those dimensions that are design-critical to the operation, manipulation, removal, or replacement task.

Access Line
A transmission path between user terminal equipment and a switching center.

Access Node
In packet switching, the switching concentration point for the transaction of subscriber's traffic to and from a network backbone system.

Accessory
(1) A basic part, subassembly, or assembly designed for use in conjunction with or to supplement another assembly, unit, or set, contributing to the effectiveness thereof without extending or varying the basic function of the assembly or set. An accessory may be used for testing, adjusting, or calibrating purposes. (2) An accessory is an assembly of a group of parts or a unit that is not always required for the operation of a set or unit as originally designed but serves to extend the function or capabilities of the set, such as headphones for a radio set supplied with a loudspeaker, a vibratory power unit for use with a set having a built-in power supply, or a remote control unit for use with a set having integral controls.

Access Phase
In an information-transfer transaction, the phase during which an individual access attempt is made. *Note*: the access phase is the first phase of an information-transfer transaction.

Access Request
A control message issued by an access originator for the purpose of initiating an access attempt.

Access Time
(1) In a telecommunication system, the elapsed time between the start of an access attempt and successful access. *Note*: access time values are measured only on access attempts that result in successful access. (2) In a computer, the time interval between the instant at which an instruction control unit initiates a call for data and the instant at which delivery of the data is completed. (3) The time interval between the instant at which storage of data is requested and the instant at which storage is started. (4) In magnetic disk devices, the time for the access arm to reach the desired track and the delay for the rotation of the disk to bring the required sector under the read-write mechanism.

Accident
An incident that results in unintended harm.

Accreditation
(1) All activities that, taken together, establish a sufficient level of confidence in the final product that the developer is able to guarantee its functional performance to specifications and to provide warranty to the customer with minimum risk of additional support at the developer's expense. (2) Accreditation is the process whereby accuracy to a predefined standard is ascertained and demonstrated for a software product. Accreditation as a term in software engineering has come to be used only recently, primarily in the DoD community to describe an authoritative endorsement of a software product. It has been used synonymously with the term *certification* in the DoD community. Accreditation requires user experience to evaluate the reliability of the product. Procedures for directly examining the software are also required before the accreditation can be made for the product in question.

Accumulator
(1) A register in which one operand can be stored and subsequently replaced by the result of the store operation. (2) A storage register. (3) A storage battery. (4) In computer applications, a device that stores a number and that, on receipt of another number, adds it to the number already stored. (5) In hydraulic and pneumatic applications, a device that stores liquid and gas under pressure for momentary use at a time when additional energy is required.

Accuracy

(1) The quality of freedom from mistake or error, that is, of conformity to truth or to a rule. [*Notes*: (a) Accuracy is distinguished from precision as in the following example: A six-place table is more precise that a four-place table. However, if there are errors in the six-place table, it may be more or less accurate than the four-place table. (b) The accuracy of an indicated or recorded value is expressed by the ratio of the error of the indicated value to the true value. It is usually expressed as a percentage. Since the true value cannot be determined exactly, the measured or calculated value of highest available accuracy is taken to be the true value or reference value. Hence, when a meter is calibrated in a given echelon, the measurement made on a meter of a higher-accuracy echelon usually will be used as the reference value. A comparison of results obtained by different measurement procedures is often useful in establishing the true value.] (2) The degree of correctness with which a measured value agrees with the true value.

Accuracy Augmentation Routine

Test routines using auxiliary test equipment that is more accurate than the automatic test equipment complement, as may be necessary when test accuracy rations cannot be met otherwise.

Accuracy Enhancement

A process that provides accuracies beyond the individual instrument capability by monitoring the instrument performance with another instrument of much greater accuracy over the duration of the test, or by means of software algorithms.

Accuracy Rating

The accuracy classification of the instruments. It is given as the limit, usually expressed as a percentage of full-scale value of reading or of programmed value, that errors will not exceed when the instrument is used under reference conditions.

Acetate Base

A photographic film support composed principally of cellulose acetate; a common type of safety film.

Acetate Film

Safety film with a base that is composed principally of cellulose acetate.

Acknowledge Character

A transmission control character transmitted by the receiving station as an affirmative response to the sending station. *Note*: an acknowledge character may also be used as an accuracy control character.

Acknowledgment

(1) A protocol data unit, or element thereof, between peer entities to indicate the status of data units that have been previously received. (2) A message from the addressee informing the originator that his communication has been received and understood.

Acoustic Coupler

A device for coupling electrical signals, by acoustical means, usually into and out of a telephone instrument. (2) A terminal device used to link data terminals and radio sets with the telephone network. *Note*: the link is achieved through acoustic (sound) signals rather than through direct electrical connection.

Acquiring Activity

The element of the agency/command that identifies and initiates a contract requirement or may have been tasked by another agency/command to be responsible for developing the contract requirement and monitoring the acquisitions. This can either be a Government or contractor flow-down to their suppliers.

Acquisition Instrument Identification Number
The Government acquiring activity's contract or purchase order number. When an order shows both a contract number and a purchase order number, the number used is determined by the acquiring activity.

Active Testing
The process of determining equipment static and dynamic characteristics by performing a series of measurements during a series of known operating conditions. Active testing may require an interruption of normal equipment operations and it involves measurements made over the range of equipment operation.

Actual Fit
The measured difference, subject to measurement uncertainty before assembly, between the sizes of two mating parts that are to be assembled.

Actual Size
The measured size of a characteristic or element subject to measurement uncertainty.

Adaptive Equalization
Equalization that is accomplished automatically while signals (user or test traffic) are being transmitted, in order to adapt to changing transmission path characteristics.

Adaptive Routing
Routing that is automatically adjusted to compensate for network changes such as traffic patterns, channel availability, or equipment failures. *Note*: the experience used for adaption comes from the traffic being carried.

Adaptive System
A system that has a means of monitoring its own performance and a means of varying its own parameters, by closed-loop action, to improve its performance.

Addendum
The addendum of an external thread is the radial distance between the major and pitch cylinders or cones, respectively. The addendum of an internal thread is the radial distance between the minor and pitch cylinders or cones, respectively. (This term applies to those threads having a recognized pitch cylinder or pitch cone.)

Adder
A device whose output data are a representation of the sum of the number represented by its input data.

Adder-Subtracter
A device that acts as an adder or subtracter, depending on the control signal received; the adder-subtracter may be constructed so as to yield a sum and a difference at the same time.

Additive Process
A process for obtaining conductive patterns by the selective deposition of conductive material on unclad base material.

Address
(1) In communications, the coded representation of the source or destination of a message.
(2) In data processing, a character or group of characters that identifies a register, a particular

part of storage, or some other data source or destination. (3) To assign to a device or item of data a label to identify its location. (4) The part of a selection signal that indicates the destination of a call. (5) To refer to a device or data item by its address.

Addressability

(1) In computer graphics, the number of addressable points on a display surface or in storage. (2) In micrographics, the number of addressable points, within a specified film frame, written as follows: the number of addressable horizontal points by the number of addressable vertical points, for example, 3,000 by 4,000.

Addressable Horizontal Positions

(1) In micrographics, the number of positions, within a specified film frame, at which a vertical line can be placed. (2) In computer graphics, the number of positions, within a specified raster, at which a full-length vertical line can be placed.

Addressable Point

Any position on a display surface that may be specified as "absolute data" in terms of the display device. Such positions are specified by "absolute coordinates," are finite in number, and form a discrete grid of the display surface.

Addressable Vertical Positions

(1) In micrographics, the number of positions, within a specified film frame, at which a horizontal line can be placed. (2) In computer graphics, the number of positions within a specified raster at which a full-length horizontal line can be placed.

Address Field

The portion of a message header that contains the destination address for the signal and the source of the signal. *Note*: in a communication network, the generally transmitted signal format contains a header, the data, and a trailer.

Address Message Sequencing

In common-channel signaling, a procedure for ensuring that addressed messages are processed in the correct order when the order in which they are received is incorrect.

Address Pattern

A prescribed structure of data used to represent the destinations of a block, message, packet, or other formalized data structure.

Adhesion

The state in which two surfaces are held together at an interface by forces or interlocking action or both.

Adhesion Promotion

The chemical process of preparing a plastic surface to provide for a uniform, well-bonded metallic over-plate.

Adhesive

A substance capable of holding two materials together by surface attachment.

Adjacent Views

Two adjoining orthographic views aligned by projectors.

Adjust

Change the value of some element of the mechanism, or the circuit of the instrument or of an auxiliary device, to bring the indication to a desired value, within a specified tolerance for a particular value of the quantity measured.

Adjustment and Calibration Time
That element of active maintenance time required to make the adjustments and/or calibrations necessary to place the item in a specified condition.

Administration
(1) Any governmental department or service responsible for discharging the obligations undertaken in the convention of the International Telecommunication Union and the Regulations. (2) The management and execution of all military matters not included in tactics and strategy, primarily in the fields of logistics and personnel management. (3) Internal management of units. (4) The management and execution of all military matters not included in strategy and tactics.

Administrative Control Number
A number assigned to one or more interchangeable purchased items by a vendor item drawing for administrative purposes. An administrative control number may also be assigned to a subcontractor designed item by a design control drawing. This number also serves as the part or identifying number for specifying such items in a parts list. The administrative control number includes the vendor item drawing or design control drawing number and is assigned in addition to the item identification assigned by the original design activity.

Administrative Time
The downtime due to non-availability of test equipment or maintenance facilities and the time due to non-availability of maintenance technicians caused by administrative functions. It is that portion of non-active maintenance time that is not included in logistic time.

Adopted Items
Items approved for inclusion in the DoD logistics system through assignment of National Stock Numbers (NSN) by the Defense Logistics Agency (DLA), or recognition by DLA of item Reference Numbers as established by manufacturer's part number, specification or drawing, or trade name (when items are identifiable by trade name only).

Advisory Signal
A signal to indicate safe or normal configuration, condition of performance, operation of essential equipment, or to attract attention and impart information for routine action purposes.

Aerial Cable
A communications cable designed for installation on, or suspension from, a pole or other overhead structure.

Aero Ballistic Missile
A wingless vehicle employing the boost-glide and continuous roll technique for flight at hypersonic speeds through the earth's atmosphere. The trajectory is ballistic to apogee, after which the vehicle assumes an angle of attack and descends, partly ballistically and partly through aerodynamic lift, to a preset altitude from which it resumes a ballistic dive to the target.

Aerodynamic Damping
Resistance to motion of a missile; caused by aerodynamic forces acting on the aerodynamic surfaces at a distance from the center of gravity incident to the pitching motion of the missile. The component of lateral velocity of such a surface in combination with the velocity due to forward speed of the missile produces an angle of attack that, in itself, provides a restoring moment.

Aerodynamic Forces
The aerodynamic effects experienced by a missile in flight, which are functions of ambient atmospheric pressure, flight Mach number, and missile size.

Aerodynamic Heating
The rise in missile-skin temperature while in flight, due to increasing air friction.

Aerodynamic Lifting Surfaces
Missile surfaces that produce normal forces to overcome gravity or to execute a maneuver. Generally of either double wedge or modified double wedge and biconvex cross-sectional profile.

Aerodynamic Loads
  - see AERODYNAMIC FORCES.

Aerodynamic Missile
A missile that uses aerodynamic forces to maintain its flight path, generally employing propulsion guidance and a winged configuration.

Aerodynamic Trajectory
A trajectory or part of a trajectory in which the missile or vehicle encounters sufficient air resistance to stabilize its flight or to modify its course significantly.

Aeronautical Earth Station
An earth station in the fixed-satellite service or, in some cases, in the aeronautical mobile-satellite service, located at a specified fixed point on land to provide a feeder link for the aeronautical mobile-satellite service.

Aero Pulse
A propulsive jet device that utilizes air to produce an intermittent thrust, as opposed to the hydro pulse, which employs water.

Aerospace, Airborne, Space
"Airborne" denotes those applications peculiar to aircraft and missile or other systems designed for operation primarily within the earth's atmosphere; "space" denotes application peculiar to spacecraft and systems designed for operation near or beyond the upper reaches of the earth's atmosphere; and "aerospace" includes both airborne and space applications.

Afocal
An optical system whose object and image point is at infinity.

Afterburning
(1) The characteristic of certain rocket motors to burn irregularly for some time after main burning and thrust have ceased. Also called *thrust trail-off*. (2) The process of fuel injection into and combustion within the exhaust jet of a turbojet engine.

Aging
(1) In a metal or alloy, a change in properties that generally occurs slowly at room temperature and more rapidly at higher temperatures. (2) The effect, on materials, of exposure to an environment for a period of time; the process of exposing materials to an environment for an interval of time.

Aided Communication
Electrically or electronically enhanced, real-time, analog or digital voice communications. In aircraft systems, the communication parts included intra-aircraft, inter-aircraft, and aircraft-to-ground links.

Air Acetylene Welding
A fuel gas welding process in which coalescence is produced by heating with a gas flame or flames obtained from the combustion of acetylene with air, without the application of pressure, and with or without the use of filler metal.

Airborne, Space, Aerospace
"Airborne" denotes those applications peculiar to aircraft and missile or other systems designed for operation primarily within the earth's atmosphere; "space" denotes application peculiar to spacecraft and systems designed for operation near or beyond the upper reaches of the earth's atmosphere; and "aerospace" includes both airborne and space applications.

Airborne Radio Relay
(1) Airborne equipment used to relay radio transmission from selected originating transmitters. (2) A technique employing aircraft fitted with radio relay stations for the purpose of increasing the range, flexibility, or physical security of communications systems.

Air Carbon Arc Cutting
An arc cutting process in which metals to be cut are melted by the head of a carbon arc and the molten metal is removed by a blast of air.

Aircraft
Any vehicle designed to be supported by air, being borne up either by the dynamic action of the air upon the surfaces of the vehicle or by its own buoyancy. The term includes fixed and movable wing airplanes, helicopters, gliders, and airships, but excludes air-launched missiles, target drones, and flying bombs.

Aircraft Station
A mobile station in the aeronautical mobile service, other than a survival craft station, located on board an aircraft.

Aircraft Station Interface
The electrical interfaces on the aircraft structure where the mission or carriage stores are electrically connected. The connection shall usually be on the aircraft side of an aircraft-to-store umbilical cable. The aircraft station interface location includes, but is not limited to, pylons, conformal and fuselage hard points, internal weapons bays, and wing tips.

Aircraft/Store Electrical Interconnection System
The AEIS is a system composed of a collection of electrical (and fiber optic) interfaces on aircraft and stores through which aircraft energize, control, and employ stores. The AEIS consists of the electrical interfaces and interrelationships between the interfaces necessary for the transfer of electrical power and data between aircraft and stores and from one store to another store via the aircraft.

Air Conditioning
In the DoD, synonym for the term "environmental control," which is the process of simultaneously controlling the temperature, relative humidity, air cleanliness, and air motion in a space to meet the requirements of the occupants, a process, or equipment.

Air Portable
Denotes materiel that is suitable for transport by an aircraft loaded internally or externally, with no more than minor dismantling and reassembling within the capabilities of user units. This term must be qualified to show the extent of air portability.

Air Sounding
Measuring atmospheric phenomena or determining atmospheric conditions, especially by means of apparatus carried by balloons, rockets, or satellites.

Air Terminal
The lightning rod or conductor placed on or above a building, structure, or external conductors for the purpose of intercepting lightning.

Alclad Sheet
Composite sheet comprised of an aluminum alloy core having on both surfaces (if on one side only, Alclad One Side Sheet) a metallurgically bonded aluminum or aluminum alloy coating that is anodic to the core, thus electrolytically protecting the core against corrosion.

ALGOL (Algorithmic Language)
A block-structured high-level programming language used to express computer programs by algorithms.

Align
To adjust circuits, equipments, or systems so that their functions are properly synchronized or their relative positions properly oriented. For example, trimmers, padders, or variable inductances in tuned circuits are adjusted to give a desired response for fixed tuned equipment or to provide tracking for tunable equipment.

Aligned Bundle
A bundle of optical fibers in which the relative spatial coordinates of each fiber are the same at the two ends of the bundle. *Note*: such a bundle may be used for the transmission of images.

Alignment
Performing the adjustments that are necessary to return an item to specified operation.

Alignment Kit
A set of instruments or tools necessary for the adjustment of electrical or mechanical components.

Alignment Mark (Printed Board)
A stylized pattern that is selectively positioned on a substrate material to assist in alignment.

Alignment Tube
A tube into which fiber ends are introduced, providing alignment prior to sealing the fiber ends in place to form a splice.

Aliquot
A small, representative portion of a larger sample.

Allocated Baseline (ABL)
(1) The initially approved documentation describing an item's functional and interface characteristics that are allocated from those of a higher-level CI, interface requirements with interfacing configuration items, additional design constraints, and the verification required to demonstrate the achievement of those specified functional and interface characteristics. (2) The approved allocated configuration documentation.

Allocated Configuration Documentation (ACD)
(1) The approved allocated baseline plus approved changes. (2) The documentation describing a CI's functional, performance, interoperability, and interface requirements that are allocated from interoperability, and interface requirements that are allocated from those of a system or higher level configuration item; interface requirements with interfacing configuration items; and the verifications required to confirm the achievement of those specified requirements.

All Over
In geometric dimensioning and tolerancing it is a symbol to indicate that a profile tolerance or other specification shall apply all over the three-dimensional profile of a part.

Allowance
The prescribed difference between the design (maximum material) size and the basic size. It is numerically equal to the absolute value of the ISO term *fundamental deviation.*

Alloy
A metallic substance, composed of two or more elements, that possesses properties different from those of its constituents.

Alloy Steel (Low-Alloy Steel)
By common custom, steel is considered to be alloy steel when the maximum range given for the content of alloying elements exceeds one or more of the following limits: manganese 1.65%, silicon 0.60%, copper 0.60%, or in which a definite range or a definite minimum quantity of any of the following elements is specified or required within the limits of the recognized field of constructional alloy steels: aluminum, boron, chromium up to 3.99%, cobalt, columbium, molybdenum, nickel, titanium, tungsten, vanadium, zirconium, or any other alloying element added to obtain a desired alloying effect. Small quantities of certain elements are present in alloy steels that are not specified or required. These elements are considered as incidental and may be present to the following maximum amounts: copper 0.35%, nickel 0.25%, chromium 0.20%, and molybdenum 0.06%.

Alphanumeric
(1) Pertaining to a character set that contains letters, digits, and sometimes other characters such as punctuation marks. (2) A character set with unique bit configurations that comprise letters of the alphabet, digits of the decimal system, punctuation symbols, and sometimes special character symbols used in grammar, business, and science.

Alphanumeric Arrangement
An ordered grouping of symbols, numbers, and letters used to form identification.

Alphanumeric Code
A character set that contains both letters and digits, special characters, and the space character.

Alpha Phase
A solid solution of one or more alloying elements in the base metal.

Altered Item
(1) An altered item is an existing item, *under the control of another design activity* or defined by a nationally recognized standardization document, that is subject to physical alteration to meet the design requirements. (2) An engineering drawing which depicts the alteration of a completed existing item from its original configuration.

Altered Item Drawing
An altered item drawing delineates the physical alteration of an existing item under the control of another design activity or defined by a nationally recognized standard. This drawing type permits the required alteration to be performed by any competent manufacturer including the original manufacturer, the altering design activity, or a third party. It establishes new item identification for the altered item.

Alternate Polarity Operation
A resistance welding process variation in which succeeding welds are made with pulses of alternating polarity.

Alternate Routing
The routing of a call or message over a substitute route when a primary route is unavailable for immediate use.

Aluminizing
The application of a film of aluminum to a surface, usually by evaporation.

Aluminum Coating
A coating composed of aluminum paste or powder and a mixing varnish or vehicle such as clear varnish or lacquer.

Aluminum Foil
A solid sheet section rolled to a thickness of less than 0.006 inches.

Ambient
The surrounding environment coming into contact with the system or component in question.

Ambient Level (Electromagnetic)
The values of radiated and conducted signal and noise existing at a specified test location and time when the test sample is not activated. Atmospheric noise and signals from man-made and natural sources all contribute to the "ambient level."

American Standard Code for Information Interchange (ASCII)
A standard seven-bit coded character set developed by the American National Standards Institute (ANSI) to be used for information interchange among information processing systems, communication systems, and associated equipment, adopted as a Federal Information Processing Standard.

American Wire Gage (AWG)
The strand system used in designating wire diameter.

Ammonia Process
Two-component diazo-type process in which both the diazo and the coupler are the base, and development is achieved by neutralizing the acidic stabilizers with vapors derived from evaporating aqua ammonia.

Ammunition
A contrivance charged with explosives, propellants, pyrotechnics, initiating composition, or nuclear, biological, or chemical material for use in connection with defense or offense including demolition, training, ceremonial, signaling, or nonoperational purposes.

Amplitude Distortion
Distortion occurring in a system, subsystem, or device when the output amplitude is not a linear function of the input amplitude under specified conditions. *Note*: amplitude distortion

is measured with the system operating under steady-state conditions with a sinusoidal input signal. When other frequencies are present, the term *amplitude* refers to that of the fundamental only.

Amplitude Equalizer
A corrective network that is designed to modify the amplitude characteristic of a circuit or system over a desired frequency range. *Note*: such devices may be fixed, manually adjustable, or automatic.

Amplitude Modulation (AM)
A form of modulation in which the amplitude of a carrier wave is varied in accordance with some characteristic of the modulating signal. *Note*: amplitude modulation implies the modulation of a coherent carrier wave by mixing it in a nonlinear device with the modulating signal, to produce discrete upper and lower sidebands, which are the sum and different frequencies of carrier and signal. The envelope of the resultant modulated wave is an analog of the modulating signal. The instantaneous value of the resultant modulated wave is the vector sum of the corresponding instantaneous values of the carrier wave, upper sideband, and lower sideband. Recovery of the modulating signal may be direct detection or by heterodyning.

Analog
Data in the form of continuously variable quantities, such as voltage, frequency, current, etc.

Analog Computer
A computer that represents variable by existing analogies. Thus, a computer that solves problems by translating existing conditions such as flow, temperature, pressure, angular position, or voltage into related mechanical or electrical equivalent quantities as an analog for the existing phenomenon being investigated. In general, it is a computer that uses an analog for each variable and produces analog as outputs. Thus, an analog computer measures continuous quantities, whereas a digital computer operates on discrete data.

Analog Data
Data represented by physical quantity that is considered to be continuously variable and whose magnitude is made directly proportional to the data or to a suitable function of the data.

Analog Decoding
A process in which an analog signal is reconstructed from a digital signal that represents the original analog signal.

Analog Encoding
Any process by which a digital signal or signals, which represent a sample or samples taken of an analog signal at a given instant or consecutive instants, are generated.

Analog Signal
(1) A signal that makes use of electrical or physical analogies (i.e., varying voltages, frequencies, distances, etc.) to produce a signal of a continuous (rather than of a pulsed or discrete) nature. (2) A nominally continuous electrical signal that varies in some direct correlation to another signal impressed on a transducer. *Note*: the electrical signal may vary its frequency phase, or amplitude, for instance, in response to changes in phenomena or characteristics such as sound, light, heat, position, or pressure.

Analog Transmission
Transmission of a continuously varying signal as opposed to transmission of a discretely varying signal.

Analytical Modeling
The technique used to express mathematically (usually by a set of equations) a representation of some real problem. Such models are valuable for abstracting the essence of the subject of inquiry. Because equations describing complex systems tend to become complicated and often impossible to formulate, it is usually necessary to make simplifying assumptions that may distort accuracy. Specific language and simulation systems may serve as aids to implementation.

Anamorphic
A term used to denote different magnification along mutually perpendicular radii. The term is also applied to an optical system that produces this condition.

Anastigmat
A lens in which the astigmatic difference is zero for at least one off-axis zone in the image plane. In such a lens, the other aberrations are sufficiently well corrected for the intended use.

Anchoring Spurs
Extensions of the lands on flexible printed wiring, extending beneath the cover layer to assist in holding the lands to the base material.

Ancillary Drawings
Ancillary drawings may be prepared to supplement end product drawings. Ancillary drawings may be required for management control, logistics purposes, configuration management, and other similar functions unique to a design activity.

Ancillary Equipment
Equipment that is auxiliary or supplementary to prime equipment installation. Ancillary equipment usually consists of standard off-the-shelf items such as an oscilloscope or distortion analyzer.

Anechoic Enclosure (Radio Frequency)
An enclosure specially designed with boundaries that absorb incident waves thereon to maintain an essentially reflection free field condition in the frequency range of interest.

Anelasticity
A characteristic exhibited by certain materials in which strain is a function of both stress and time such that, while no permanent deformations are involved, a finite time is required to establish equilibrium between stress and strain in both the loading and unloading directions.

Anerobic
Having an absence of oxygen in the uncombined state.

Angle Modulation
Modulation in which the phase angle or frequency of a sine wave carrier is varied.

Angle of Incidence
The angle between the line of direction of anything (as a ray of light or line of sight) striking a surface and a line perpendicular to that surface drawn to the point of contact.

Angularity
Angularity is the condition of a surface or axis at a specified angle (other than 90°) from a datum plane or axis.

Angular Misalignment
Angular departure of one fiber from the axis defined by the other when two fibers are connected or spliced.

Anisotropic
A term used to denote a substance that exhibits different properties when tested along axes in different directions. Not isotropic; having mechanical and/or physical properties that vary with direction relative to natural reference axes inherent in the material.

Annealing
Heating to and holding at a suitable temperature and then cooling at a suitable rate, for such purposes as reducing hardness, improving machinability, facilitating cold working, producing a desired microstructure, or obtaining desired mechanical, physical, or other properties.

Annular Ring
That portion of conductive material completely surrounding a hole.

Anode
(1) An electrode at which oxidation of the anode surface or some component of the solution is occurring (opposite of cathode). (2) The positive pole in an electric arc.

Anodic Inhibitor
A chemical substance or combination of substances that prevent or reduce the rate of the anodic or oxidation reaction by a physical, physical-chemical, or chemical action.

Anodic Polarization
A polarization of the anode: the anode potential becomes more noble (positive) because of the nonreversible conditions resulting when corrosion current flows.

Anodic Protection
A technique to reduce corrosion of a metal surface under some conditions by passing sufficient anodic current to it to cause its electrode potential to enter and remain the passive region.

Anodizing
Electrolytic process forming an oxide coating, such as on aluminum, for corrosion and wear resistance. Differs from electroplating, as the work is done by the anode instead of the cathode.

Anolyte
The electrolyte of an electrolytic cell adjacent to the anode.

Anomalies
Deviations from normal conditions; in communications terminology, the effect on radio paths from changing conditions in transmission media.

Antenna Effective Area (in a Given Direction)
The ratio of the power available at the terminals of an antenna to the incident power density of a plane wave from that direction, polarized coincident with the polarization that the antenna would radiate.

Antenna Effective Length
The ratio of the antenna open circuit induced voltage to the intensity of the field component being measured.

Antenna Factor
That factor that, when properly applied to the meter reading of the measuring instrument, yields the electric field strengths in volts/meter or the magnetic field strength in amperes/meter. *Note*: this factor includes the effects of antenna effective length as well as mismatch and transmission loss.

Antenna Gain (Relative)
The ratio of the power required at the input of a reference antenna to the power supplied to the input of the given antenna to produce, in a given direction, the same field at the same distance. When not otherwise specified, the gain figure for an antenna refers to the gain in the direction of the radiation main lobe. *Note*: in applications using scattering modes of propagation the full gain of an antenna may not be realized in practice, and the apparent gain may vary with time.

Antenna, Phased Array
An array antenna whose beam direction or radiating pattern is controlled primarily by the relative phases of the excitation coefficients of the radiating elements.

Anthropometric Dimensions
Measured dimensions that describe the size and shape of the human body. These dimensions are often presented in the form of summary statistics that describe the range of body dimensions that are observed in a population.

Anthropometry
The scientific measurement and collection of data about human physical characteristics and the application (engineering anthropometry) of these data to the design and evaluation of systems, equipment, and facilities.

Anti-fouling
A chemical control of a surface to prevent the attachment or growth of marine organisms when submerged. The toxicity can be invoked by careful selection of the metal itself, by chemical treatments, or by suitable coatings.

Anti-interference
Pertaining to equipment, processes, or techniques used to reduce the effect of natural and man-made noise on radio communications.

Anti-jam
Pertaining to equipment, processes, or techniques used to reduce the effect of jamming on a desired signal.

Anti-node
A point in a standing (i.e., stationary) wave at which the amplitude is a maximum; i.e., there is a crest. *Note*: the type of wave should be identified, such as a voltage wave or a current wave.

Antioxidant
A substance that prevents or reduces the rate of oxidation due to exposure of material to air or oxygen.

Antireflection Coating
A thin, dielectric or metallic film (or several such films) applied to an optical surface to reduce its reflectance and thereby increase the transmittance. *Note*: the ideal value of the refractive index of a single-layer film is the square root of the product of the refractive indices on either side of the film, the ideal optical thickness being one quarter of a wavelength.

Antistatic Property
This term refers to the reduction of triboelectric charge generation. Antistatic materials minimize the generation of static charges. This property is not dependent upon material resistivity.

Aperture
(1) The rectangular hole portion of an aperture, camera, copy, or image card. (2) An opening in a shield through which electromagnetic energy passes. (3) An opening or hole through which light or matter may pass. In an optical system, it is equal to the diameter of the largest entering beam of light that can travel completely through the system. This may or may not be equal to the aperture of the objective.

Aperture Card
An unprocessed tabulating card that contains an aperture (rectangular hole) specifically designed for the subsequent insertion of a developed frame of silver halide camera microfilm.

Aperture, Clear (CA)
The opening in the mount of an optical system or any component thereof that limits the extent of the bundle of rays' incident on the specific surface. It is usually circular and specified by its diameter. Clear aperture is sometimes referred as *free aperture* or *objective aperture*.

Aperture, Effective
Equivalent to the diameter of the largest bundle of rays that can be imaged by the optical system.

Aperture, Free
 - see APERTURE, CLEAR.

Aperture, Objective
 - see APERTURE, CLEAR.

Application Certification
A formal statement by a mutually approved and authorized agency that an application has successfully undergone some conformance test procedures.

Application Data
The next assembly and the model number, nomenclature, or equivalent designator of the assembled unit of which a part or assembly is a component.

Application Identifier (AI)
The field of two or more characters at the beginning of an Element String that uniquely defines its format and meaning.

Approval (Drawing)
An endorsement applied manually or electronically attesting to the correctness of a document or a revision made on a document.

Approval Indicator (Drawing)
Any symbol adopted by the design activity to indicate approval.

Approval/Contractual Implementation
The acceptance by the Government of a document as complete and suitable for its intended use. Approval/Contractual implementation of configuration documentation means that the approved documentation is subject to the Government's configuration control procedures.

Approved Data
The configuration management controlled master version of data formally submitted to and approved by the Government.

Approved Equipment Test Procedure
The test procedure furnished by the Government or furnished by a contractor in accordance with the requirements of the contract or order and approved by the Government in accordance with the contract or order.

Approved Item Name
An approved item name is a name approved by the Directorate of Cataloging, Defense Logistic Services Center and published in the *Cataloging Handbook H6*, Federal Item Name Directory for Supply Cataloging.

Aqueous
Pertaining to water: an aqueous solution is made by using water as a solvent.

Aqueous Cleaning Solution
An aqueous cleaning solution is a water based cleaner whose constituents are soluble inorganic compounds such as silicates or phosphates or soluble organic compounds such as nonionic surfactant or combinations thereof. Examples of aqueous cleaning solutions include commercial detergents which generally contain both soluble inorganic and organic compounds and Navy oxygen cleaner (NOC) which contains only soluble inorganic compounds.

Aramid
A manufactured fiber in which the fiber-forming substance consists of a long-chain synthetic aromatic polyamide in which at least 85% of the amide linkages are attached directly to two aromatic rings.

Arc
A discharge of electricity across a gap in a circuit or between two electrodes.

Arc Brazing
A brazing process in which the heat required is obtained from an electric arc.

Arc Cutting
A group of cutting processes that melt the metals to be cut with the heat of an arc between an electrode and the base metal.

Arc Length
In geometric dimensioning and tolerancing it is a symbol that indicates that a dimension is an arc length measured on a curved outline.

Arc Resistance
The resistance of a material to the effects of a high-voltage, low-current arc (under prescribed conditions) passing across the surface of the material. The resistance is stated as a measure of total elapsed time required to form a conductive path on the surface (material carbonized by the arc).

Arc Seam Welding
An arc-welding process wherein coalescence at the faying surfaces is produced continuously by heating with an electric arc between an electrode and the work. The weld is made without preparing a hole in either member. Filler metal or a shielding gas or flux may or may not be used.

Arc Spot Welding
An arc-welding process wherein coalescence at the faying surfaces is produced in one spot by heating with an electric arc between an electrode and the work. The weld is made without preparing a hole in either member. Filler metal or a shielding gas or flux may or may not be used.

Arc Strike
An arc strike is any inadvertent change in the characteristics of the finish weld or adjacent base material resulting from an arc of heat generated by the passage of electrical energy between the surface of the finished weld or base material and a current source, such as welding electrodes or magnetic particle inspection prods.

Arc Welding
A group of welding processes in which fusion is obtained by heating with an electric are or arcs, with or without the application of pressure and with or without the use of filler metal.

Armed Component
A weapon component is *armed* when it has responded to an influence to change its state from Safe or Disabled (inoperable), to Arm, or Enabled (operable).

Armor Piercing (AP) Munitions
Armor piercing munitions are munitions such as projectiles and bombs designed to penetrate armor and other resistant targets by means of kinetic energy and the physical characteristics of the warhead.

Arms Export-Control Act (AECA)
The law set out in 22 U.S.C. 2751–2794. This requires obtaining an approval from the Department of State for exporting defense articles and services, including technical data related to munitions and military equipment. It is implemented by the International Traffic in arms Regulations (ITAR) set out in 22 CFR 121–130.

Arm-to-Arm
As used in the ordnance of guided missiles, the changing from a safe condition to a state of readiness in preparation for initiation or ignition of the propulsion system.

Arrangement Drawing
An arrangement drawing depicts the physical relationship of significant items using appropriate projections or perspective views. Reference dimensions may be included. An arrangement drawing does not establish item identifications.

Array
(1) An arrangement of elements in one or more dimensions. (2) In a programming language, an aggregate that consists of data objects with identical attributes, each of which may be uniquely referenced by subscription.

Array Processor
A processor capable of executing instructions in which the operands may be arrays rather than data elements.

Article
A nonspecific term used to denote any unit or product including materials, parts, assemblies, equipment, accessories, and computer software.

Articulation Index
A weighted number representing, for a given set of speech and noise conditions, the effective proportion of the normal speech signal that is available to a listener for conveying speech intelligibility. Articulation index is computed from acoustical measurements (or estimates) of the speech spectrum and of the effective masking spectrum, and is defined on a scale of 0–1.0.

Artificial Intelligence
A field aimed at pursing the possibility that a computer can be made to behave in a manner that humans recognize as intelligent behavior to each other.

Artificial Weathering
Exposure to laboratory conditions that may be cyclic, involving changes in temperature, relative humidity, radiant energy, and any other elements found in the atmosphere in various geographical areas.

Artwork
An accurately scaled configuration that is used to produce the Artwork Master or Production Master.

Artwork Master
An accurately-scaled configuration that is used to produce the Production Master.

As Applicable
This term is intended to require inclusion of those data elements necessary to establish the engineering definition or end product requirements.

ASCII
American Standard Code for Information Interchange, used extensively in data transmission. The code includes 128 upper- and lower-case letters, numbers, and special-purpose symbols, each encoded by a unique seven-bit binary number.

ASCII Test
A subset of the ASCII, common to all computer devices, consisting principally of the printable characters.

Aspect Ratio
In an essentially two-dimensional rectangular structure (e.g., a panel), the ratio of the long dimension to the short dimension. However, in compression loading, it is sometimes considered to be the ratio of the load direction dimension to the transverse dimension. Also, in fiber micromechanics, it is referred to as the ratio of length to diameter.

Aspherical
A term used to characterize a departure from the spherical shape.

Assemble
To translate a set of some language statements into a simple form, usually the machine code of a particular machine. The translation is often a one-to-one transformation.

Assembler
(1) A computer program used to assemble. Synonymous with *assembly* program. (2) A tool that translates programs written in symbolic machine language into actual machine language programs.

Assembly
(1) A number of parts or subassemblies or any combination thereof joined together to perform a specific function, and subject to disassembly without degradation of any of the parts. (2) A number of parts or subassemblies or any combination thereof joined together to perform a specific function and capable of disassembly (e.g., an audio frequency amplifier). The distinction between an assembly and subassembly is determined by the individual application. An assembly in one instance may be a subassembly in another where it forms a portion of an assembly.

Assembly Drawing
An assembly drawing defines the configuration and contents of the assembly or assemblies depicted thereon. It establishes item identification for each assembly. Where an assembly drawing contains detailed requirements for one or more parts used in the assembly, it is a detail assembly drawing.

Assembly Language
A low-level programming language that is actually a symbolic machine language.

Associated Data
Any document referenced on an engineering drawing that establishes some portion of the engineering requirements.

Associated Detail Specification
The associated detail specification is an extension of a general specification that covers detailed requirements for specific parts, materials, or equipments.

Associated List
A tabulation of engineering information pertaining to an item depicted on an engineering drawing, or by a set of drawings (e.g., parts list, data list, and index list).

A-Stage
An early stage in the reaction of thermosetting resins in which the material is still soluble in certain liquids and may be liquid or capable or becoming liquid upon heating.

Atomic Hydrogen Welding
An arc welding process that produces coalescence of metals by heating them with an electric arc maintained between two metal electrodes in an atmosphere of hydrogen. Shielding is obtained from the hydrogen. Pressure may or may not be used, and filler metal may or may not be used.

Attachment
A basic part, subassembly, or assembly designed for use in conjunction with another assembly, unit, or set, contributing to the effectiveness thereof by extending or varying the basic function of the assembly, unit, or set.

Attenuation
A general term used to denote a decrease in magnitude (of power or field strength) in transmission from one point to another. It may be expressed as a ratio or, by extension of the term, in decibels.

Attribute
(1) A characteristic or property that is appraised in terms of whether it does or does not exist, (e.g., go or no-go) with respect to a given requirement. (2) An element of a constituent of a document that has a name and a value and that expresses a characteristic of this constituent or a relationship with one or more constituents. (3) Nondimensional thread element and characteristic taken singly or in a group. Inspection/evaluation by limit gages is an attribute inspection.

Audit
To conduct an independent review and examination of system records and activities in order to test the adequacy and effectiveness of data security and data integrity procedures, to ensure compliance with established policy and operational procedures, and to recommend any necessary changes.

Audit Review File
A file created by executing statements included in a program for the explicit purpose of providing data for auditing.

Audit Trail
(1) A record of both completed and attempted accesses and service. (2) Data in the form of a logical path linking a sequence of events, used to trace the transactions that have affected the contents of a record. (3) A chronological record of system activities that is sufficient to enable the reconstruction, review, and examination of the sequence of environments and activities surrounding or leading to an operation, a procedure, or an event in a transaction from its inception to final results. *Note*: *audit trail* may apply to information in an automated information system, to the routing of messages in a communications system, or to material exchange transactions.

Austenitic
An adjective describing a face-centered cubic crystal structure found in ferrous materials, usually at elevated temperatures; however, certain stainless steels (300 series) can exhibit this structure at ambient temperatures.

Authenticate
(1) To establish, usually by challenge and response, that a transmission attempt is authorized and valid. (2) To verify the identity of a user, device, or other entity in a computer system or to verify the integrity of data that have been stored, transmitted, or otherwise exposed to possible unauthorized modification. (3) A challenge given by voice or electrical means to attest to the authenticity of a message or transmission.

Authentication
(1) A security measure designed to protect a communications system against acceptance of a fraudulent transmission or simulation by establishing the validity of a transmission, message, or originator. (2) A means of identifying individuals and verifying their eligibility to receive specific categories of information. (3) Evidence by proper signature or seal that a document is genuine and official. (4) A security measure designed to protect a communication system against fraudulent transmissions.

Authorization
(1) The rights granted to a user to access, read, modify, insert, or delete certain data, or to execute certain programs. (2) The granting of access rights to a user, a program, or a process.

Autoclave
A closed vessel for producing an environment of fluid pressure, with or without heat, to an enclosed object that is undergoing a chemical reaction or other operation.

Autoclave Molding
A process similar to the pressure bag technique. The layup is covered by a pressure bag, and the entire assembly is placed in an autoclave capable of providing heat and pressure for curing the part. The pressure bag is normally vented to the outside.

Automata
The class of sequential machines that, by alteration of internal state, are capable of performing logical computational or repetitive routines, i.e., automatic processors, computers, decoders, controllers, and their associated equipment.

Automated Component Insertion
Automated component insertion is the act or operation of assembling individual components to printed boards by means of computer-controlled component-insertion equipment.

Automated Inspection Equipment
Measuring or gaging instruments that make a pass or fail determination of a dimensional characteristic without human interaction. Automated inspection equipment is usually integral to a production line and is often computer controlled.

Automated Path Analysis
A software technique that scans source code in order to design an optional set of test cases to exercise the primary paths in a software module.

Automated Radio
A radio with capability for automatically controlled operation by electronic devices that require little or no operator intervention.

Automatic
Pertaining to a process or device that, under specified conditions, functions without intervention by a human operator.

Automatic Data Processing (ADP)
(1) An interfacing assembly of procedures, processes, methods, personnel, and equipment to perform automatically a series of data processing operations that result in a change in the semantic content of the data. (2) Data processing, by means of one or more devices, (a) that uses common storage for all or part of a computer program, and also for all or part of the data necessary for execution of the program; (b) that execute user-written or user designated programs; (c) that perform user designated symbol manipulation, such as arithmetic operations, logic operations, or character-string manipulations; and (d) that can execute programs that modify themselves during their execution. Automatic data processing may be performed by a standalone unit or by several connected units. (3) Data processing largely performed by automatic means. (4) That branch of science and technology concerned with methods and techniques relating to data processing largely performed by automatic means.

Automatic Gain Control
A process or means by which gain is automatically adjusted in a specified manner as a function of input level or another specified parameter.

Automatic Test Equipment (ATE)
(1) Test, measurement, and diagnostic equipment that performs a program to test functional or static parameters, to evaluate the degree of performance degradation, or to perform fault isolation of unit malfunctions. The decision making, control, or evaluative functions are conducted with minimal reliance on human intervention. (2) An equipment that is designed to automatically conduct analysis of functional or static parameters, evaluate the degree of performance degradation, and perform isolation of item malfunctions.

Automatic Testing
The process by which the localization of faults, possible prediction of failure, or validation that the equipment is operating satisfactorily is determined by a device that is programmed to perform a series of self-sequencing test measurements without the necessity of human direction after its operations have been initiated.

Automation
(1) The implementation of processes by automatic means. (2) The investigation, design, development, and application of methods of rendering processes automatic, self-moving, or self-controlling. (3) The conversion of a procedure, a process, or equipment to automatic operation.

Auxiliary Equipment
Equipment that is supplementary to a prime equipment installation. Auxiliary equipment usually consists of standard off-the-shelf items such as oscilloscopes and distortion analyzer. Also called *ancillary equipment*.

Auxiliary Heating or Cooling
External heating or cooling devices not normally part of the equipment configuration.

Auxiliary Operation
An off-line operation performed by equipment not under control of the processing unit.

Auxiliary Power
An alternate source of electric power serving as backup for the primary power at the station main bus or prescribed sub-bus. *Note*: an off-line unit provides electrical isolation between the primary power and the critical technical load; an online unit does not. Class A power source is a primary power source, i.e., a source that assures an essentially continuous supply of power. Types of auxiliary power service include Class B: a standby power plant to cover extended outages (days); Class C: a quick-start (10–60 s) unit to cover short-term outages (hours); and Class D: an uninterruptible (no-break) unit using stored energy to provide continuous power within specified voltage and frequency tolerances.

Auxiliary View
Auxiliary views are used to show true shape and relationship of features that are not parallel to any of the principal planes of projection.

Availability
A measure of the degree to which an item is in operable and committable state at the start of the mission, when the mission is called for at an unknown (random) point in time.

Availability (Achieved)
The percentage of time the system is operating when considering only operating time and total maintenance (scheduled and unscheduled) time.

Availability (Inherent)
The percentage of time the system is operating when considering only operational time and unscheduled (corrective) maintenance time.

Availability Model
A model (or models) that predicts the expected ratio of system up-time to the total operating time.

Availability (Operational)
A measure of the degree to which an item is either operating or is capable of operating at any random point in time when used in a typical maintenance and supply environment.

Avalanche Photodiode (APD)
A photodetector used in high-speed (broad bandwidth) lightwave systems. The avalanche feature results from the rush of electrons across a junction under a very high reverse bias. The APD requires a much higher reverse bias and has a much higher cutoff frequency than a PIN photodiode and is therefore more sensitive at high frequencies.

Average Outgoing Quality (AOQ)
For a particular process average, the AOQ is the average quality of outgoing product, including all accepted lots or batches plus all rejected lots or batches after the rejected lost or batches have been effectively 100% inspected and all defectives replaced by nondefectives.

Avionics
Application of electronics to aviation and astronautics.

Avionics Applications
Software engineering applied to flight control systems for aircraft.

Axis, Cylinder
The meridian perpendicular to that in which the cylindrical power functions.

Axis of Thread
The axis of the thread pitch cylinder or cone.

Axis of a Weld
A line through the length of a weld perpendicular to the cross-section at its center of gravity.

Axis, Optical
The line formed by the coinciding principal axis of a series of optical elements comprising an optical system. It is the line passing through the centers of curvatures of the optical surfaces. The optical centerline.

Axis, Principal
A straight line connecting the centers of curvature of the refracting surfaces of a lens. In a mechanical sense, a line joining the centers of a lens as it is placed in a mount. The principal axis is the optical axis of a lens.

Axis, Secondary
A line formed by the chief ray of an oblique bundle of rays.

Axis, Visual
An imaginary line from the object through the nodal point of the eye to the fovea, or point of sharpest retinal acuity.

Axonometric Projection
An axonometric projection is one in which he projectors are perpendicular to the plane of projection and parallel to each other. The principle surfaces and edges of a cube or other rectangular object are all inclined to the plane of projection. The angles between the principle edges, or axes, of the object shall not be 90° on the drawing. The relationship between the three angles shall be such that the mutual perpendicularity of the axes on the object is maintained. Axonometric is divided into isometric, diametric, and trimetric projections.

Azimuth, Angle of
The angle measured clockwise in a horizontal plane, usually from north (may be true north, Y-north, grid north, or magnetic north).

B

Backplane
An interconnection device having terminals (such as solderless wrapped connections) on one side and usually having connector receptacles on the other side; used to provide point-to-point electrical interconnections. The point-to-point electrical interconnections may be printed wiring.

Back Spall
Refers to fragments of metal displaced from the back of steel armor as a result of ballistic attack, normally from an area of greater size than the projectile caliber.

Back Trace
The process of tracing back from the failure site to the primary inputs and making an input assignment so that a distinguishable test can be produced at the failure site.

Backup
A capability that returns a user to the last previous display in a defined transaction sequence. Also refers to the practice of preserving a second copy of files for data protection purposes.

Backup File
A copy of a file made for purposes of later reconstruction of the file, if necessary.

Back Weld
A weld deposited at the back of a single-groove weld.

Bag Molding
A method of molding or laminating that involves the application of fluid pressure to a flexible material that transmits the pressure to the material being molded or bonded. Fluid pressure usually is applied by means of air, steam, water, or vacuum.

Balance
To adjust the impedance of circuits and balance networks to achieve specified return loss objectives at junctions of two-wire and four-wire circuits.

Balanced Line
A line or circuit using two conductors instead of one conductor and ground (common conductor). The two sides of the line are symmetrical with respect to ground. Line potentials to ground and line currents are equal but of opposite phase at corresponding points along the line.

Ballistic Coefficient
The numerical measure of the ability of a missile to overcome air resistance. It is dependent upon the mass, the diameter, and the configuration.

Ballistic Protection System
A ballistic protection system represents some combination of one or more elements made of basic armor materials or composites (in some cases supplemented by non-armor materials) to form an effective ballistic protection device. This includes installation equipment and hardware.

R. Hanifan, *Concise Dictionary of Engineering: A Guide to the Language of Engineering*, 29
DOI 10.1007/978-3-319-07839-7_2, © Springer International Publishing Switzerland 2014

Ballistics
The science or art that deals with the motion, behavior, appearance, or modification of missiles or other vehicles acted upon by propellants, wind, gravity, temperature, or any other modifying substance, condition, or force.

Band Frequency
In communications and electronics, a continuous range of frequencies extending between two limiting frequencies.

Bandpass
Number of hertz expressing the difference between the limiting frequencies at which the desired fraction (usually half power) of the maximum output is obtained.

Bandpass Filter
A device that passes all frequencies within its designed ranges and bars passage to all frequencies not within the ranges.

Bandpass Limiter
A device that imposes hard limiting on a signal and contains a filter that suppresses the unwanted products (harmonics) of the limiting process.

Band-Stop Filter
A device that bars passage of frequencies within its designed range and allows passage of higher or lower frequencies, or both.

Bandwidth
(1) The difference between the limiting frequencies within which performance of a device, in respect to some characteristic, falls within specified limits. (2) The difference between the limiting frequencies of a continuous frequency band.

Bang-Bang Control
A control method wherein the corrective control applied to a missile always is applied to the full extent of the servo motion arc.

Bar
(1) The darker element of a bar code. (2) A solid wrought product that is long in relation to its cross section, which is square or rectangular (excluding plate and flattened wire) with sharp or rounded corners or edges, or is a regular hexagon or octagon, and in which at least one perpendicular distance between parallel faces is 0.375 inch or greater.

Bar Code
An array of rectangular bars and spaces in a predetermined pattern.

Bar Width
The perpendicular distances across a bar measured from a point on one edge to a point on the opposite edge. Each point will be defined as having a reflectance that is 50% of the difference between the background and bar reflectance.

Baseband
(1) The spectral band occupied by an unmodulated signal. *Note:* Baseband transmission is usually characterized by being much lower in frequency than the signal that results if the baseband signal is used to modulate a carrier or subcarrier. (2) In facsimile, the frequency of a signal equal in width to that between zero frequency and maximum keying frequency.

Baseline
A configuration identification document or a set of such documents formally designated by the Government at a specific time during a Configuration Item (CI) life cycle. Baselines, plus approved changes from those baselines, constitute the current approved configuration identification. For configuration management purposes, there are three baselines, which are established sequentially, as follows: Functional Baseline, Allocated Baseline, and Product Baseline.

Baseline Program
A program possessing well defined capabilities and functions that are decreed to be the starting point for further program development.

Base Material
The insulating material upon which a conductor pattern may be formed. The base material may be rigid or flexible. It may be a dielectric sheet or insulated metal sheet.

Base Material Thickness
The thickness of the base material excluding metal foil or material deposited on the surfaces.

Base Metal
The metal to be welded.

Basic Dimension
A numerical value used to describe the theoretically exact size, profile, orientation, or location of a feature or datum target. It is the basis from which permissible variations are established by tolerances on other dimensions, in notes, or in feature control frames.

Basic Grid
 - see GRID.

Basic Item
A term used to distinguish an end item of equipment from individual components, assemblies, subassemblies, and parts; e.g., overhaul of the basic item weapon system (AH-64 helicopter airframe including components repaired while on the airframe) versus overhaul of engines, accessories, and components, and assemblies that have been removed from the basic item and overhauled to meet established supply requirements for the helicopter.

Basic Part
See also Part. One piece or two or more pieces joined together that are not normally subject to disassembly without destruction of designed use.

Basic Reference Designation
The simplest form of a reference designation, consisting only of a class letter portion and a number (namely, without mention of the item within which the reference designated item is located).

Basic Reference Standards
Those standards used to maintain physical and electrical units in the laboratory, and which serve as the starting point of the chain of sequential measurements carried out in the laboratory.

Basic Shaft System
A system of fits in which the design of the shaft is the basic size, and the allowance, if any, is applied to the hole.

Basic Statistical Methods
Applies the theory of variation through the use of basic problem-solving techniques and statistical process control, and includes control chart construction and interpretation (for both variables and attributes data) and capability analysis.

Basic Statistics
Applies the theory of variation through the use of applied statistics in collecting and summarizing data, analyzes grouped as well as unwrapped data, and includes histogram construction and interpretation.

Basic Value
Attribute value that is unconditionally allowed in document interchange in the context of a given application profile.

Batch
 - see LOT.

Batch Processing
(1) The processing of data or the accomplishment of jobs accumulated in advance in such a manner that the user cannot further influence the processing while it is in progress. (2) The processing of data accumulated over a period of time. (3) Loosely, the execution of computer programs serially. (4) Pertaining to the technique of executing a set of computer programs such that each is completed before the next program of the set is started. (5) Pertaining to the sequential input of computer programs or data.

Battle-Short
Battle-short is a function that disables equipment protection and personnel safety interlocks in order to keep the equipment online during high readiness states. It maintains the maximum available mission readiness and system availability by avoiding interlock caused shutdowns and prolonged start-ups.

Battle-Short Switch
A switch used on high-priority equipment designed to bypass or short-circuit interlock switches or devices during emergency conditions.

Bayesian Model
Indexing term. Refers to the mathematical methodology used to construct, or which is the form assumed by, a particular model.

Bayonet Coupling
A quick-coupling device for plug and receptacle connectors, accomplished by rotation of a cam-operating device designed to bring the connector halves together.

Bearer Bar
A rectangular bar pattern circumscribing the bar code, often used with a bar code directly printed on corrugated fiberboard.

Bed-of-Nails Fixture
A test fixture consisting of a frame and a holder containing a field of spring-loaded pins that make electrical contact with a planar test object.

Bed-of-Nails Interface Adapter
A type of interface adapter that uses a series of pogo-pins or nails to make contact with the unit under test.

Begin-End Block
Begin-end block is a collection of computer program statements bracketed by begin and end statements. The latter delimits the scope of names and is also activated by normal sequential flow of control. The term came from the ALGOL 60 programming language. These statements are often used to define the limits of a collection of codes such as a module or subroutine.

Behavioral Model
A mathematical function that relates cause and effect quantitatively.

Behavior Modeling
Describing what a component of a software system will do in terms of an abstraction of the component's operation, which focuses upon effect rather than cause.

Bellows Contact
A connector contact that is a flat spring folder used to provide a uniform spring rate over the full tolerance range of the mating unit.

Below Deck Areas
An area in ships that is surrounded by a metallic structure such as the hull or superstructure of metallic surface ships, the hull of a submarine, the screened areas or rooms of non-metallic ships, the screened areas of ships utilizing a combination of metallic/non-metallic material for hull and superstructure or a deck mounted metallic shelter.

Benchmark
A test point for comparison purposes; in microprocessor-based equipment, a benchmark program is one used to compare aspects of performance among competing systems.

Bend Test
A test of ductility by bending or folding, usually with steadily applied forces. In some instances, the test may involve blows to a specimen having a cross section that is essentially uniform over a length several times as great as the largest dimension of the cross section.

Best Operating Capability
The upper level maintainability value estimated to be technically feasible within the stated time frame and within reasonable cost constraints.

Best Wire Size
For symmetrical threads, the size of a wire that would touch at the pitch diameter on a basic profile thread of zero lead angle. For nonsymmetrical threads, the best wire size will contact the load flank at a point twice the distance above the pitch line that the contact point on the clearance flank is below the pitch line.

Between
In geometric dimensioning and tolerancing it is a symbol used to indicate that a tolerance or other specification apply across multiple features or to a limited segment of a feature between designated extremities.

Bevel
A type of edge preparation.

Bevel Angle
The angle formed between the prepared edge of a member and a plane perpendicular to the surface of the member.

Beveling
A type of chamfering.

Bias
A systemic deviation of a value from a reference value. (2) The amount by which the average of a set of values departs from a reference value. (3) An electrical, mechanical, magnetic, or other force field applied to a device to establish a reference level to operate the device. (4) Effect on telegraph signals produced by the electrical characteristics of the terminal equipment.

Bias Distortion
Distortion affecting a two-condition (binary) coding in which all the significant intervals corresponding to one of the two significant conditions have uniformly longer or shorter durations than the corresponding theoretical durations. *Note:* the magnitude of the distortion is expressed in percent of a perfect unit pulse length.

Bidirectional Bus
A conductor or group of conductors that transmits and receives digital data on the same line.

Bidirectional Code
A bar code format that permits reading in complementary (opposite) directions across the bars and spaces.

Bifurcated Contact
A connector contact (usually a flat spring), which is slotted lengthwise to provide additional, independently operated points of contact.

Bilateral Tolerance
A tolerance in which variation is permitted in both directions from the specified dimension.

Binary
Pertaining to a characteristic or property involving a selection, choice, or condition in which there are two possibilities.

Binary Code
A code that makes use of exactly two distinct characters, usually 0 and 1.

Binary Digit (BIT)
(1) A character used to represent one of the two digits in the numeration system with a base of two, each digit representing one of two, and only two, possible states of a physical entity or system. (2) In binary notation, either of the characters 0 or 1. (3) A unit of information equal to one binary decision or the designation of one of two possible and equally likely states of anything used to store or convey information.

Binder
A bonding resin used to hold strands together in a mat or preform during manufacture of a molded object.

Binomial Random Variable
The number of successes in independent trials where the probability of success is the same for each trial.

Bipolar Signal
A signal having two polarities, both of which are not zero. Usually, bipolar signals are symmetrical with respect to zero amplitude.

Birefringence
The difference between the two principal refractive indices (of a fiber) or the ratio between the retardation and thickness of a material at a given point.

Bit
(1) A contraction of the term *binary digit* and hence either a 0 or a 1 in the base-two number system. (2) A character used to represent one of the two digits in the numeration system with a base of two, each digit representing one of two, and only two, possible states of a physical entity or system. (3) In binary notation, either of physical entity or system. (4) A unit of information equal to one binary decision or the designation of one of two possible and equally likely states of anything used to store or convey information.

Bit Map
(1) Individual bits within a message or field that provides a specific definition. (2) A two or three dimensional data field representing a pel array.

Blackbody
A body that absorbs all the radiant energy that strikes it; a perfect radiator and a perfect absorber. It is a contraction of the term *ideal blackbody* and is often used synonymously with *ideal radiator, full radiator* or *complete radiator.*

Black Box
An actual or a conceptual device that transforms input data into output data according to a prescribed functional relationship, but whose internal mechanization is not necessarily known.

Black Crest Thread
A thread whose crest displays the unfinished cast, rolled, or forged surface.

Blank
(1) An unprocessed or partially processed piece of base material, or metal-clad base material, cut from a sheet or panel and having the rough dimensions of a printed board. (2) A pressed-glass mold with the approximate size and shape of the optical element to be ground and polished.

Bleeder Cloth
A nonstructural layer of material used in the manufacture of composite parts to allow the escape of excess gas and resin during cure. The bleeder clots is removed after the curing process and is not part of the final composite.

Bleeding
A condition in which a plated hole discharges process material or solution from crevices or voids.

Blister
A localized swelling and separation between any of the layers of a laminated base material, or between base material and conductive foil. (It is a form of delamination.)

Block
A basic layout object that corresponds to a rectangular area within a frame and consisting of only a single type of content, i.e., raster graphics content.

Block Change Concept
For hardware configuration items, an engineering change implementation concept that desig-nates a number (i.e., a block) of consecutive production units of the configuration item to

have an identical configuration on delivery and in operation. (Using this concept, the production run is divided into "blocks" of units. The production line incorporation point for a proposed ECP is delayed to coincide with the first unit of the next block, or retrofit is required at least for all already delivered units of the current block.) For computer software configuration items, once the product baseline has been established, the concept required the accumulation and the simultaneous implementation of a number of routine software changes to minimize the number of interim versions and related documentation.

Block Diagram
A diagram of a system, instrument, or computer in which the principal parts are represented by suitably associated geometrical figures to show both the basic functions and the functional relationships among the parts.

Block-Structured Language
A higher-order programming language that demarcates related sequences of code, or blocks, usually with the statements *begin* and *end*.

Blow-Back (Microfilm)
To enlarge or make an enlargement of an image. Also the copy prepared by this method.

Blowup
A photomechanical line reproduction enlarged from a proof of a smaller-sized halftone, particularly a newspaper (coarse screen) halftone showing highlight effects and accentuated contrast. The term applied to a photographic enlargement.

Blue Line
When a cut for a pipe or fitting nears the inner wall, the outer surface turns blue which is referred to as cutting to the blue line. The blue line results from oxidation caused by the reduced ability of the thinned pipe or fitting to dissipate heat generated by the power tool cutting the surface.

Blueprint
A blue-background print with white lines made on iron-sensitized paper by printing through transparent positive copy.

Blueprint Process
Reproduction method using light-sensitive iron salts, which produces a negative blue image from a positive master.

Bluing
Subjecting the scale-free surface of a ferrous alloy to the action of air, steam, or other agents at a suitable temperature, thus forming a thin blue film of oxide and improving the appearance and resistance to corrosion.

Blunt Start Thread
This term designates the removal of the incomplete thread at the starting end of the thread. This is a feature of threaded parts that are repeatedly assembled by hand, such as hose couplings and thread plug gages, to prevent cutting of hands and crossing threads.

Board Thickness (Printed Wiring Board)
The overall printed wiring board thickness includes metallic depositions, fusing, and solder resist. The overall thickness is measured across the printed wiring board extremities (thickest part), unless a critical area, such an edge-board contacts or card guide mounting location, is identified on the master drawing.

Boattail
The conical section of a ballistic body that progressively decreases in diameter toward the tail to reduce overall aerodynamic drag.

Bolt, Clevis
An externally threaded fastener whose threaded and unthreaded portions are of one nominal diameter and are separated by a narrow circumferential groove. The head has a recess for holding or driving.

Bolt, Close Tolerance
An externally threaded fastener whose unthreaded portion is of a specified grip length and is machined to a tolerance of .001 inch or less. Items over 1.000 inch in diameter shall have a tolerance of .0015 inch or less. The nominal major diameter of the threads shall be at least .001 inch below the minimum shank diameter, but not below the minimum major diameter for applicable class of fit. The head is designed for external wrenching. The minimum tensile strength shall be less than 160,000 pounds per square inch.

Bolt, Eye
An externally threaded device whose threaded portion is of one nominal diameter, without a head, but with the unthreaded end either bent more than 225 degrees or cast, forged, or punched to resemble an eye.

Bolt, Hook
An externally threaded device whose threaded portion is of one nominal diameter, without a head but with the unthreaded end bent not over 225 degrees.

Bolt, Lag
An externally threaded fastener having a square or hexagon head and with a continuous thread (wood screw type or fetter drive type) extending from a gimlet or cone point for a distance of slightly more than one-half the length of the bolt.

Bolt, Machine
An externally threaded fastener whose threaded and unthreaded portions are each of one nominal diameter .190 inch or larger. The length of the unthreaded portion (of hexagon head fasteners) is controlled and is machined to a tolerance greater than that specified for a bolt, close tolerance. The head is designed for external wrenching only.

Bolt, Self-Locking
A Bolt, machine or screw, cap, hexagon head with the added characteristic of a locking feature incorporated in the design of the head or in the threads.

Bolt, Shear
A bolt, close tolerance except that item is fabricated from material having a minimum tensile strength of 160,000 pounds per square inch or greater.

Bolt, Shoulder
A bolt, machine or screw, cap hexagon head that has a round unthreaded neck or shank, all or part of which is of greater diameter than the threaded portion.

Bolt, Square Neck
A headed externally threaded fastener whose threaded portion has a square neck directly beneath the head.

Bolt, Tee Head
An externally threaded fastener whose threaded portion is of one nominal diameter, and with a head specifically designed to fit in a slot and hold against turning.

Bolt, U
An externally threaded fastener bent approximately 180 degrees in the shape of the letter U and with both ends threaded.

Bond
(1) The electrical connection between two metallic surfaces established to provide a low-resistance path between them. (2) Any fixed union existing between two objects that result in electrical conductivity between the two objects. Such union occurs either from physical contact between conducting surfaces of the objects or from the addition of a firm electrical connection between them.

Bond Direct
An electrical connection utilizing continuous metal-to-metal contact between the members being joined.

Bond Indirect
An electrical connection employing an intermediate electrical conductor between the bonded members.

Bonding
The process of establishing the required degree of electrical continuity between the conductive surfaces to be joined.

Bonding Layer
An adhesive layer used in bonding together other discrete layers of a multilayer printed board during lamination.

Bond Line
The cross section of the interface between thermal spray deposited and substrate or the interface between adhesive and adherent in an adhesive bonded joint.

Bond Permanent
A bond not expected to require disassembly for operational or maintenance purposes.

Bond Semipermanent
Bonds expected to require periodic disassembly for maintenance, or system modification, and that can be reassembled to continue to provide a low-resistance interconnection.

Bond Strength
The force per unit area required to separate two adjacent layers of a board by a force perpendicular to the board surface (see Peel Strength).

Bootstrap
(1) A technique or device designed to bring about a desired state by means of its own action. (2) That part of a computer program that may be used to establish another version of the computer program. (3) The automatic procedure whereby the basic operating system of a processor is reloaded following a complete shutdown or loss of memory.

Bore Sight
To adjust the line of sight of the sighting instrument of a weapon parallel to the axis of the bore. Also applied to the process of aligning other equipment, such as radar mounts, directors, etc. As a noun, the term defines an optical instrument for checking alignment.

Bottom-Up Design
The design of the system starting with the lowest-level routines and proceeding to the higher-level routines that use the lower levels.

Bottom-Up Implementation
The implementation of the system starting with the lowest-level routines and proceeding to the higher-level routines that use the lower levels.

Boundary, Inner
A worst-case boundary (that is, locus) generated by the smallest feature (MMC for an internal feature and LMC for an external feature) minus the stated geometric tolerance and any additional geometric tolerance (if applicable) from the feature's departure from its specified material condition.

Boundary, Least Material (LMB)
The limit defined by a tolerance or combination of tolerances that exists on or inside the material of a feature.

Boundary, Maximum Material (MMB)
The limit defined by a tolerance or combination of tolerances that exists on or outside the material of a feature.

Boundary, Outer
A worst-case boundary (that is, locus) generated by the largest feature (LMC for an internal feature and MMC for an external feature) plus the geometric tolerance and any additional geometric tolerance (if applicable) from the feature's departure from it specified material condition.

Bow
The deviation from flatness of a board characterized by a roughly cylindrical or spherical curvature such that, if the board is rectangular, its four corners are in the same plane.

Braze
A weld wherein coalescence is produced by heating to suitable temperatures above 800 degrees F, and by using a nonferrous filler metal, having a melting point below that of the base metals. The filler metal is distributed between the closely fitted surfaces of the joint by capillary attraction.

Brazement
An assembly whose component parts are joined by brazing.

Braze Welding
A method of welding whereby a groove, fillet, plug or slot weld is made using a nonferrous filler metal, having a melting point below that of the base metals, but above 800 degrees F. The filler metal is not distributed in the joint by capillary attraction.

Brazing
A group of welding processes in which fusion is obtained by heating to suitable temperatures above 800°F but below the melting temperature of the base metal and adding a nonferrous filler metal with a melting point below that of the base metals. The filler metal may be deposited in an open joint between the parts or may be distributed between closely abutting surfaces of the joint by capillary attraction.

Brazing Sheet
Sheet of a brazing alloy, or sheet clad with a brazing alloy on one or both sides.

Breadboard
An assembly of circuits or parts used to prove the feasibility of a device, circuit, system, or principle with little or no regard to the final configuration or packaging of the parts.

Bridge
A functional unit that interconnects two local area networks that use the same logical link control procedure but may use different medium access control procedures.

Bridge Fault
Short circuits or leakage between adjacent paths (lands and traces) on a printed circuit board.

Bridging, Electrical
The formation of a conductive path between conductors.

Bright Annealing
Annealing in a protective medium to prevent discoloration of the bright surface.

Broadgoods
A term loosely applied to prepreg material greater than about 12 inches in width, usually furnished by suppliers in continuous rolls. The term is currently used to designate both collimated uniaxial tape and woven fabric prepregs.

Browsing
The act of searching through automated information system storage to locate or acquire information without necessarily knowing of the existence or the format of the information being sought.

B-Stage
(1) An intermediate stage in the polymerization reaction of certain thermosetting resins; the state in which most prepregs are stored and shipped. (2) An intermediate stage in the reaction of a thermosetting resin in which the material softens when heated and swells when in contact with certain liquids but does not entirely fuse or dissolve. Materials are usually procured to this stage to facilitate handling and processing prior to final cure.

Buckles
Buckles are indentations in the surface of sand casting that result from expansion of the sand.

Buffer
An isolating circuit used to avoid reaction of a driven circuit on the corresponding driving circuit.

Buffer Material
A resilient material used to protect crack-sensitive components from excessive stresses generated by the conformal coating.

Bug
(1) One or more software bugs exist in a system if a software change is required to correct a single major or minor error so as to meet specified or implied system performance requirements. (2) A concealed microphone or listening device or other audio surveillance device. (3) A mistake in a computer program. (4) To install means for audio surveillance. (5) A semiautomatic telegraph key. (6) A mistake or malfunction.

Bug Seeding/Tagging
The process of adding bugs (or errors) to those already assumed to be in a program for the purpose of obtaining an estimate for the number of natural bugs remaining in the program. It is also assumed that ratio of the number of undiscovered seeded bugs to the total number of

bugs seeded can serve as an indication of the degree of "debuggedness" or reliability of the program.

Building
The fixed or transportable structure that houses personnel and equipment and provides the degree of environmental protection required for reliable performance of the equipment housed within.

Building Block
Generation of a program as an isolated building block. Necessary independent subprograms are generated first, followed by generation of the dependent functions.

Built-In Flexibility
Built-in flexibility is the ability of a system to immediately handle different logical situations. Built-in flexibility increases system complexity proportionately. In a well designed system, the initial measure of built-in flexibility will be almost equal to the complexity measure.

Built-In Test (BIT)
(1) An integral capability of the mission system or equipment that provides an automated test capability to detect, diagnose, or isolate failures. (2) A test approach using built-in test equipment or self-test hardware and software that is internally designed into the supported end item to test all or a part of that item.

Built-In Test Equipment (BITE)
(1) Hardware that is identifiable as performing the built-in test function; a subset of BIT. (2) Any identifiable device that is a part of the supported end item and is used for testing that supported end item.

Bulge
A bulge is a swelling of a printed wiring board, usually caused by internal delamination or separation of fibers.

Bulk Items
Bulk items are those constituents of an assembly or part (such as oil, wax, solder, cement, ink, damping fluid, grease flux, welding rod, twine, or chain) for which the quantity is not readily predeterminable, the physical nature of the material is such that it is not adaptable to pictorial representation, or the finished size is obtainable through use of such tools as shears, pliers, or knives without further machining operation, and the final configuration is such that it can be described in writing without the necessity of pictorial representation.

Bulk Materials
  - see BULK ITEMS.

Bundle
A group of optical fibers or conductors associated together and usually in a single sheath.

Bung Hole
An opening in a barrel or drum through which liquids can be poured to fill or to empty it. In fiber drums, the bung hole is in the head. In wooden barrels, it is usually in the sidewall. In steel drums, there may be one, two, or three openings, one or two being in the head while one may be in the sidewall. The largest opening is called the *bung*.

Burn-In
To operate electronic items under specified environmental and test conditions to eliminate early failures and to stabilize the items prior to actual use.

Burnout
The point in time or in the missile trajectory when combustion of fuels in the rocket engine is terminated by other than programmed cutoff.

Burst
A single transmission, over the communications channel, that consists of start-of-message, a header, and operational data packets.

Burst Synchronization
The maintenance of burst count integrity between transmitting and receiving data controllers.

Bus
A conductor, or group of conductors, which serve as the path for carrying digital control, address, and information signals. Also power distribution between controlling and controlled electronic items.

Bus Bar
A conduit, such as a conductor on a printed board, for distributing electrical energy.

Butter Coat
A commonly used term to describe a higher-than usual surface resin.

Buttering
A surfacing variation in which one or more layers of weld metals are deposited on the groove face of one member (for example, a high alloy weld deposit on steel base metal that is to be welded to a dissimilar base metal). The buttering provides a suitable transition on weld deposit for subsequent completion of the butt joint.

Butt Joint
A weld between two members lying approximately in the same plane.

Butt Weld
An erroneous term for a weld in a butt joint. See Butt Joint.

Bypass
(1) Broadly, the use of any telecommunication facilities or services in circumvention of the local exchange carrier. *Note:* the alternative facilities or services may be either customer provided or vendor supplied. (2) Generally, an alternate circuit around some user equipment, group of equipments, or system element. *Note:* it is usually provided to allow system operation to continue when the bypassed portion is inoperable.

Bypassable Interlock
A bypassable interlock is an automatic switch with a manually operated electrical bypass device to allow equipment maintenance operations on energized equipment

Byte
(1) Unit of measurement used in determining data file size or available space for data storage. One byte is approximately equivalent to one character of type. (2) A string of bits whose length is the smallest accessible as a unit in a computer memory; also, the length used to represent a character.

Byte Boundary
A position in a binary data stream where, if the stream were packed into bytes, an integer number of completely filled bytes would result.

C

Cabinet

A protection housing or covering for two or more units or pieces of equipments. A cabinet may consist of an enclosed rack with hinged doors.

Cable

(1) An assembly of one or more conductors or optical fibers, or a combination of both, within an enveloping sheath. (2) A message sent by cable, or by extension, any means of telegraphy. (3) Two or more insulated conductors, solid or stranded, contained in a common covering, or two or more insulated conductors twisted or molded together without common covering, or one insulated conductor with a metallic covering shield or outer conductor.

Cable Assembly Drawing

A cable assembly drawing depicts an electrical cable assembly of defined length and establishes item identification for that assembly.

Cabinet Projection

A cabinet projection is an oblique projection in which the projectors make an angle with the plane of projection, which reduces distance along or parallel to the receding axis to one-half of that for cavalier projections.

Cache

Processor caches store data from (slower) main memory on special chip cache memory, where it can be accessed and reused much more efficiently. A cache is used to speed up data transfer and may be either temporary or permanent. Memory and disk caches are in every computer to speed up instruction execution and data retrieval and updating. These temporary caches serve as staging areas, and their contents are constantly changing.

Cage Code
- see COMMERCIAL AND GOVERNMENT ENTITY CODE

Calibration

(1) Comparison of a measurement standard or instrument of unknown accuracy with another standard or instrument of known accuracy to detect, correlate, report, or eliminate by adjustment any variation in the accuracy of the unknown standard or instrument. (2) Those measurement services provided by designated depot and/or laboratory facility teams, who by the comparison of two instruments, one of which is a certified standard of known accuracy, detect and adjust any discrepancy in the accuracy of the instrument being compared with the certified standard.

Camber

The planar deflection of a flat cable or flexible laminate from a straight line of specified length.

Camera Card

An unprocessed tabulating card whose aperture contains undeveloped silver halide camera microfilm.

Canard Configuration
A missile airframe that has the horizontal surfaces used for trim and control on the forward portion of the body, and the main lifting surfaces attached to the rear section. The forward control surfaces can be used to vary lift by varying the angle of attack of the body-wing combination.

Cancel
A capability that regenerates or re-initializes the current display without processing or retaining any changes made by the user.

Canceled Drawing
A drawing which has been removed from the drawing system and the part or assembly shown on the drawing is removed from all next assembly usage. Drawings which have been superseded or become obsolete are also considered to be canceled drawings.

Candela
A unit of luminous intensity defined such that the luminance of a blackbody radiator at the temperature of solidification of platinum is 60 candelas per square centimeter.

Candlepower
A unit of measure of the illuminating power of any light source. The number of candles in the luminous intensity of a source of light. A luminous intensity of one candle produces one lumen of luminous flux per steradian of solid angle measured from the source.

Cannibalization
The removal of serviceable items from a piece of equipment to repair another.

Cannot Duplicate
A fault indicated by BIT or other monitoring circuitry that cannot be confirmed at the first level of maintenance.

Capability
(1) A measure of the ability of an item to achieve mission objectives given that the item performs as specified through the mission. (2) A capability is defined as an abstract encapsulation of the data needed to define access to a protected object with respect to security. (3) Capabilities are discretely identified elements of performance that are expected (either formally or informally) of a product or combination of products. A failure is the absence of one or more capabilities during the use of a product. Severity of a failure is directly proportional to the value of the absent capabilities to the user. An error becomes a failure when software is incapable of reestablishing its capabilities in an error environment.

Capillary Action
The force by which liquid, in contact with a solid, is distributed between closely fitted faying surfaces of the joint to be brazed or soldered.

Captcha
A category of technologies used to ensure that a human is making an online transaction rather than a computer. Random words or letters are displayed in a camouflaged and distorted fashion so that they can be deciphered by people, but not by software. Users are asked to type in the text they see to verify they are human.

Captive Flight
A missile flight test in which the missile, or components of it, are carried on an aircraft as a means of simulating flight conditions.

Captive Test
The firing of a missile's rocket motor while the vehicle is mounted on a test stand, performed to check the performance of the system. Flight conditions cannot be simulated, but engines usually can be operated up to full thrust.

Carbon-Arc Welding
An arc-welding process in which fusion is obtained by heating with an electric arc between a carbon electrode and the work. No shielding is used, and pressure and filler metal may or may not be used.

Carbon Black
A black pigment produced by the incomplete burning of natural gas or oil. It is widely used as a filler or pigment, particularly in the rubber industry. Because it possesses ultraviolet protective properties, it is used in polyethylene systems such as cold-water piping and black agricultural sheet.

Carbon Precipitate
A precipitate composed of a compound of carbon with one or more metallic elements, often found in ferrous materials but also in aluminum and titanium.

Carbon Steel
Carbon steel is classed as such when non-minimum content is specified or guaranteed for aluminum, chromium, columbium, molybdenum, nickel, titanium, tungsten, vanadium, or zirconium; when the minimum for copper does not exceed 0.40%; or when the maximum content specified or guaranteed for any of the following elements does not exceed the percentages noted: manganese, 1.65; silicon, 0.60; copper, 0.60.

Card Tester
An instrument for testing and diagnosing printed circuit cards.

Carriage Store Interface
The electrical interface on the carriage store structure where the aircraft is electrically connected. This connection is usually on the store side of an aircraft-to-store umbilical cable.

Carriage Stores
Suspension and release equipment that is mounted on aircraft on a nonpermanent basis as a store are classified as a carriage store. Pylons and primary racks are not considered carriage stores.

Carriage Store Station Interface
The electrical interfaces on the carriage store structure where the mission stores are electrically connected. This connection is usually on the carriage store side of an umbilical cable.

Carrier
(1) A wave suitable for modulation by an information-bearing signal to be transmitted over a communication system. (2) An unmodulated emission. *Note:* the carrier is usually a sinusoidal wave or a recurring series of pulses.

Carrier Dropout
A short-duration loss of carrier signal.

Carrier Frequency
(1) The frequency of a carrier wave. (2) The frequency of an unmodulated wave capable of being modulated or impressed with a second (information-carrying) signal. *Note:* in frequency modulation, the carrier frequency is also referred to as the *center frequency.*

Carrier Noise Level
The noise level resulting from undesired variations of a carrier in the absence of any intended modulation.

Carrier Sense
In a local area network, an ongoing activity of a data station to detect whether another station is transmitting.

Carrier Synchronization
In a radio receiver, the generation of a reference carrier with a phase closely matching that of a received signal.

Carrier System
A multichannel telecommunications arrangement wherein a number of individual data and/ or voice circuits are multiplexed for transmission between nodes of a network with de-multiplexing occurring as required.

Case Hardened
Subjected to a surface heat-treating process that produces a highly stressed surface.

Case Hardening
Hardening a ferrous alloy so that the outer portion or case is made substantially harder than the inner portion or core.

Case Temperature
The case temperature is the temperature (typically the hottest temperature point found on the mounting surface of the device) measured on the external surface of the device's package.

Casting
An object of finished or near-finished shape obtained by filling a mold with molten metal and allowing it to solidify.

Catastrophic Failure
A failure mode measured is both sudden and complete. This failure causes cessation of one or more fundamental functions.

Catastrophic Fault
(1) A physical condition causing a catastrophic failure. (2) A fault measured will destroy the system or subsystem and its function almost immediately.

Cathode
(1) The electrode of an electrolytic cell at which reduction occurs (opposite of *anode*). (2) The negative pole in an electric arc.

Cathode Ray Tube (CRT)
An electronic tube in which a well defined and controllable beam of electrons is produced and directed to a surface to give a visible or otherwise detectable display or effect. The face of a CRT is used in some interactive graphic display devices as the display surface.

Cathodic Polarization
A polarization of the cathode: The cathode potential becomes more active (negative) because of the nonreversible conditions resulting when a corrosion current flows.

Cation
An ion having a positive charge that is attracted to the cathode (opposite of *anion*).

Caul Plates
Smooth metal plates, free of surface defects, the same size and shape as a composite lay-up, used immediately in contact with the lay-up during the curing process to transmit normal pressure and to provide a smooth surface on the finished laminate.

Cause and Effect Diagram
A simple tool for individual or group problem-solving that uses a graphic description of the various process elements to analyze potential sources of process variation. Also called *fish-bone diagram* (after its appearance) or *Ishikawa diagram* (after its developer).

Caustic Embrittlement
A form of stress-corrosion cracking occurring in steel exposed to alkaline solutions.

Caution Signal
A signal that alerts the operator to an impending dangerous condition requiring attention, but not necessarily immediate action.

Cavalier Projection
A cavalier projection is an oblique projection on which the projectors make 45 degrees with the plane of projection.

Cavitation
A formation and sudden collapse of vapor bubbles in a liquid that momentarily can cause localized regions of high pressure.

Cavitation Corrosion
Damage of a material associated with the collapse of cavities in the liquid at a solid-liquid interface.

CDRL
 - see CONTRACT DATA REQUIREMENTS LIST.

Cellophane
Transparent film made of regenerated cellulose. Cellophane is inherently greaseproof and, by suitable coating, may be made moisture proof and heat sealable.

Cellular Rubber
Cellular rubber is defined as a product containing cells or small, hollow receptacles. The cells may be either open and interconnected or closed and not interconnected. The cells should be uniform and free of large voids or seams.

Celluloid
A plastic substance, highly flammable and possessing a distinct odor, prepared from nitrocellulose and camphor and formed into sheets, rods, and tubes for further manufacture of novelties, spectacle frames, drawing instruments, etc.

Cellulose
A carbohydrate constituent of the walls and skeletons of vegetable cells.

Cellulose Acetate
A thermoplastic material made of cellulose with acetic anhydride and acetic acid.

Cellulose Acetate-Butyrate
A thermoplastic material that can be converted into shapes or film similar to cellulose acetate. Has property of toughness with relatively high impact strength and shock resistance.

Center

A collection of units and items in one location, which provides facilities for the administrative control in an area of responsibility which is specifically assigned for development and maintenance of installations, control of personnel, or conduct of tactical operations.

Center Frequency

(1) In frequency modulation, the rest frequency (frequency of the carrier before modulation). (2) In facsimile, the frequency midway between the picture-black and picture-white frequencies.

Central

A grouping of sets, units or combinations thereof operated conjunctively in the same locations for a common specific function. It may provide facilities for controlling switching, monitoring, etc., electronic and electrical equipment from one central location.

Central Processing Unit (CPU)

(1) The portion of a computer that includes circuits controlling the interpretation and execution of instructions. (2) The portion of a computer that executes programmed instructions, performs arithmetic and logical functions on data, and controls input/output functions.

Certificate of Compliance

A document signed by an authorized party affirming that the supplier of a product or service has met the requirement of the relevant specifications, contract, or regulation.

Certificate of Conformance

A contractor's written statement, when authorized by contract, certifying that supplies or services comply with contract requirements.

Certification

(1) Verification that a support test system is capable, at the time of certification demonstration, of correctly assessing the quality of the items to be tested. This verification is based on an evaluation of all support test system elements and establishment of acceptable correlation among similar test systems. (2) A process, which may be incremental, by which a contractor provides objective evidence to the contracting agency that an item satisfies its specified requirements.

Chalking

A development of a loose, removable powder at or just beneath an organic coating surface.

Chamfer

(1) A broken corner used to eliminate an otherwise sharp edge. (2) The conical surface at the starting end of a thread.

Change Control Board (CCB)

A Change Control Board is an organization established by the contractor for managing and control of product changes.

Channel

(1) A connection between initiating and terminating nodes of a circuit. (2) A single path provided from a transmission medium either by physical separation, e.g., multipair cable, or by electrical separation, e.g., frequency, or time-division multiplexing. (3) A single unidirectional or bidirectional path for transmitting or receiving (or both) electrical or electromagnetic signals, usually in distinction from other parallel paths. (4) Used in conjunction with a predetermined letter, number, or codeword to reference a specific radio frequency. (5) A path along

which signals can be sent; e.g., data channel, output channel. (6) The portion of a storage medium that is accessible to a given reading or writing station; e.g., track, band. (7) In information theory, that part of a communications system that connects the message source with the message sink.

Character
Letter, digit, or other special form that is used as part of the organization, control, or representation of data. A character is often in the form of a spatial arrangement of adjacent or connected strokes.

Character Code
A correspondence between a character set and a set of integers.

Characteristic, Critical
A critical characteristic is one that analysis indicates is likely, if defective, to create or increase a hazard to human safety or to result in failure of a weapon system or major system to perform a required mission.

Characteristic Impedance
The ratio of voltage to current in a propagating wave, i.e., the impedance offered to this wave at any point of the line. (In printed wiring, its value depends on the width of the conductor, the distance from the conductor to ground planes, and the dielectric constant of the media between them.)

Characteristic, Major
A major characteristic is one that analysis indicates is not critical but is likely, if defective, to result in failure of an end item to perform a required mission.

Characteristic, Minor
A minor characteristic is one that analysis indicates is significant to product quality but is not likely, if defective, to impair the mission performance of the item.

Character Set
A set of graphic symbols independent of font. A character set does not include specification of codes to represent characters.

Characters per Inch (CPI)
The number of bar coded characters that are displayed in each inch of bar code.

Chassis
The metal structure that supports the electrical components that make up the unit or system.

Chassis Wiring
Chassis wiring may consist of hookup wire, lead wire, shielded cable, jacketed multiconductor cable, coaxial cable, or twisted multiconductor groups of wires or cables, or wires and cables used to connect electrical or electronic elements within the same equipment.

Check
A process for determining accuracy.

Check Bit
A binary digit used for error detection; for example, a parity bit.

Check Character
A single character, derived from and appended to a data item, that can be used to detect errors in processing or transmitting a data item.

Check Digit
A single digit, derived from and appended to a data item, that can be used to detect errors in processing or transmitting a data item.

Checkout
A man/machine task to determine that the equipment is operating satisfactorily and is ready for return to service.

Checkpoint
A place in a routine where a check, or a recording of data for restart purposes, is performed.

Checksum
(1) The sum of every byte contained in an input/output record or memory used for assuring integrity of the programmed entry. (2) The sum of a group of data items that is stored with the group and is used for checking purposes. The data items are either numeric or may be treated as numeric for the purposes of calculating the checksum. (3) An error detection technique based on a summation operation performed on the bits to be checked.

Chemical Agent Resistant Coating (CARC)
A chemical agent resistant coating enhances the decontamination process for combat and support equipment that is subjected to surface contamination by chemical attack on the battlefield. Chemical agents deposited on the surface of CARC paints remain on the surface and can be removed with decontaminant procedures without destroying the coating.

Chemical Conversion Coating
A protective or decorative coating produced by a chemical reaction of a metal with a chosen environment.

Chemical Fuel
A fuel that requires an oxidizing agent to produce combustion and the resultant thrust in a liquid or solid propellant rocket engine. Chemical fuels undergo reactions or rearrangement of atoms to form different molecules.

Chemical Hole Cleaning
The chemical process for cleaning conducive surfaces exposed within a hole (see Etchback).

Chemical Pressurization
The use of high-pressure gases, produced by combing a fuel and an oxidizer or from the decomposition of a substance, to pressurize propellant tanks in a rocket vehicle.

Chemical Vapor Deposition Technique
In optical fiber manufacturing, a process in which deposits are produced by heterogeneous gas-solid and gas-liquid chemical reactions at the surface of a substrate. *Note:* this method is often used in fabricating optical fiber preforms by causing gaseous materials to react and deposit glass oxides. The preform may be processed further in preparation for pulling into an optical fiber.

Chemical Warfare Agent
A chemical agent in a solid, liquid, or gas that produces lethal or damaging effects on man, animals, plants, or materials, or negatively affects their performance.

Chuffing
The irregular or intermittent burning of propellants in a liquid rocket engine that creates a low-frequency pressure oscillation parallel to the thrust axis and sometimes produces an irregular puffing noise. Also called *chugging, combustion resonance,* or *pogo effect.*

Cipher
Any cryptographic system in which arbitrary symbols or groups of symbols represent units of plain text of regular length, usually single letters, or in which units of plain text are rearranged, or both, in accordance with certain predetermined rules.

Circuit
(1) An electronic closed-loop path between two or more points used for signal transfer. (2) One complete traverse of the fiber feed mechanism of a winding machine; one complete traverse of a winding band from one arbitrary point along the winding path to another point on a plane through the starting point and perpendicular to the axis. (3) The complete path between two terminals over which one-way or two-way communications may be provided. (4) An electronic path between two or more points, capable of providing a number of channels. (5) A number of conductors connected together for the purpose of carrying an electrical current.

Circuit Breaker
A protective device for opening and closing a circuit between separable contacts under both normal and abnormal conditions. *Note:* circuit breakers may be of many types and sizes, and they are usually classified according to the medium in which the interruption takes place, e.g., oil (or other liquid) or air (or other gas).

Circuit Card Tester
An instrument for testing and diagnosing printed circuit cards.

Circuit Malfunction Analysis
The logical, systematic examination of circuits and their diagrams to identify and analyze the probability and consequence of potential malfunctions for determining related maintenance or maintainability design requirements.

Circuit Noise Level
At any point in a transmission system, the ratio of the circuit noise at that point to some arbitrary amount of circuit noise chosen as a reference.

Circuitry
A complex of circuits describing interconnection within or between systems.

Circuitry Layer
A circuitry layer is a layer of a printed wiring board containing conductors. It also includes both ground planes and voltage planes.

Circuit Simulator
A computer program that simulates the operation of an electronic circuit. In digital applications, it also analyzes the efficiency of stimulus test patterns on itself.

Circularity (Roundness)
Circularity is a condition of a surface of revolution where (a) for a cylinder or cone, all points of the surface intersected by any plane perpendicular to a common axis are equidistant from that axis or (b) for a sphere, all points of the surface intersected by any plane passing through a common center are equidistant from that center.

Circular Mil
A unit of area equal to the area of a circle whose diameter is 1 mil (0.001 inch); equal to square mil × 0.78540. Used chiefly in specifying cross-sectional areas of round conductors.

Circumferential Separation
(a) crack in the plating extending around the entire circumference of a plated-through hole, (b) in the solder fillet around the lead wire, (c) in the solder filler around an eyelet, or (d) at the interface between a solder fillet and a land.

Clad
A condition of the base material when a relatively thin layer or a sheet of metal foil has been bonded to one or both of its sides; e.g., a metal-clad base material.

Cladding
(1) When referring to an optical fiber, a layer of material of lower refractive index, in intimate contact with a core material of higher refractive index. (2) When referring to a metallic cable, a process of covering with a metal (usually achieved by pressure rolling, extruding, drawing, or swaging) until a bond is achieved.

Clarity
Code possesses the characteristic clarity to the extent that it is concise, straightforward (lack of tricky, obscure code), and understandable; has clear control structure and uniform style; is self-contained with respect to documentation; and makes appropriate use of macros and change levels.

Class
This term provides additional categorization of differences in characteristics other than those afforded by type classification, which does not constitute a difference in quality or grade, but are for specific, equally important uses. It is usually designated by Arabic numerals; thus, "class 1" and "class 2."

Classification Code
Those codes assigned in the process of classification of characteristics.

Classification of Characteristics
The process of assigning classification codes (Critical, Major, Minor) to design characteristics of an item.

Classification of Defects
A classification of defects is the enumeration of possible defects of the unit of product classified according to their seriousness.

Classified Information Processing System (CLIPS)
Equipment, device, or system that is electrically powered and that processes, converts, reproduces, or otherwise manipulates any form of classified information.

Class of Thread
An alphanumerical designation to indicate the standard grade of tolerance and allowance specified for a thread.

Clean Room
A room in which the concentration of airborne particles is controlled and that contains one or more clean zones.

Clean Zone
A defined space in which the concentration of airborne particles is controlled to meet a specified airborne particulate cleanliness class.

Clearance Fit
A clearance fit is one having limits of size so prescribed that a clearance always results when mating parts are assembled.

Clearance Hole
A hole in the conductive pattern larger than, but coaxial with, a hole in the printed board base material.

Clear Aperture
  - see APERTURE, CLEAR.

Clinched Leads
Component leads that extend through the printed board and are formed to effect a spring-action, metal-to metal electrical contact with the conductive pattern prior to soldering.

Clinched-Wire Through Connection
A connection made by a wire that is passed through a hole in a printed board and subsequently formed, or clinched, in contact with the conductive pattern on each side of the board, then soldered.

Clip
Same as scissor.

Clips
Small metal pieces placed between the disc and the major to achieve proper fusing.

Clock
(1) A device that generates periodic signals used for synchronization. (2) A reference source of timing information. (3) A device providing signals used in a transmission system to control the timing of certain functions such as the duration of signal elements or the sampling rate. (4) A device that generates periodic, accurately spaced signals used for such purposes as timing, regulation of the operations of a processor, or generation of interrupts. (5) In geometric dimensioning and tolerancing, a term used to define angular orientation.

Closed Cell (Expanded)
Closed cell is a product usually made by subjecting a rubber compound to a gas, such as nitrogen, under high pressure. It may also be made by incorporating gas-forming materials in the compound.

Closed Circuit
(1) In radio and television transmission, used to indicate that the programs are transmitted directly to specific users and not broadcast for general consumption. (2) In telecommunications, a circuit dedicated to specific users. *Note:* the circuit may be active or inactive at any given time. (3) A completed electrical circuit.

Closed Loop Testing
Testing in which the input stimulus is controlled by the equipment output monitor.

Cloud
A wide area network (WAN) commonly is depicted as a cloud, which serves to obscure its complex inner workings from view. Data just pops in on one side of the cloud land pops out on the other side.

Coalescence
The growing together or growth into one body of the materials being welded.

Coat, Hard
A term applied to the process, or to the result of the process, of producing (usually) dielectric coatings that are more durable under adverse conditions than those produced from other processes.

Coated Thread
A thread with one or more applications of additive material. This includes dry film lubricants, but excludes soft or liquid lubricants that are readily displaced in assembly and gaging. Plating and anodizing are included as coatings.

Coating, Antireflection
A class of single or multilayer coatings that are applied to a surface or surfaces of a substrate for the purpose of decreasing the reflectance of the surface and increasing the transmission of the substrate over a specified wavelength range.

Coating, High-Reflecting
A broad class of single- or multilayer coatings that are applied to a surface for the purpose of increasing its reflectance over a specified range of wavelengths. Single films of aluminum or silver are common; but multilayers of at least two dielectrics are utilized when low absorption is imperative.

Coatings, Protective
Films that are applied to a coated or uncoated optical surface primarily for protecting this surface from mechanical abrasion, chemical corrosion, or both. An important class of protective coatings consists of evaporated thin films of titanium dioxide, silicon monoxide, or magnesium fluoride. For example, a thin layer of silicon monoxide may be added to protect an aluminized surface.

Coat, Soft
A term designating the soft coating applied to coated optics to differentiate between the harder and more durable coating known as *hard coat*. Certain evaporating coatings are not capable of forming a hard coat and are easily removed by cleaning. Cryolite is a soft coat material.

Coaxiality
Coaxiality is the condition in which the axes of two or more surfaces of revolution are coincident.

Cocked Head
The seating of either the manufactured head or the upset head at an angle other than perpendicular to the shank of the rivet parallel to the surface.

Cocuring
The act of curing a composite laminate and simultaneously bonding it to some other prepared surface during the same cure cycle.

Codec
(1) In communications engineering, it s used in reference to integrated circuits, or chips that perform data conversion. In this context, the term is an acronym for "coder/decoder." This type of codec combines analog-to-digital conversion and digital-to-analog conversion functions in a single chip. (2) The term is also an acronym that stands for "compression/decompression." A codec is an algorithm, or specialized computer program, that reduces the number of bytes consumed by large files and programs.

Code Density
The number of characters that can appear per unit of length, normally expressed in characters per inch (CPI).

Coding
That part of the test program development process where the test sequences are translated into the language of the ATE controller.

Coefficient of Linear Thermal Expansion
The change in length per unit length resulting from a one degree rise in temperature.

Coefficient of Variation
The ratio of the population (or sample) standard deviation to the population (or sample) mean.

Coils, Active
The number of coils used in computing the total deflection of a spring.

Coils, Total
The number of active coils plus the coils forming the ends (compression springs).

Coincidence
Agreement as to position. In a coincidence rangefinder, the two half images of a distant object are in "coincidence" when they are in exact alignment.

Cold Cracks
Cracks that appear on radiograph as a straight line, usually continuous throughout its length, and generally exist singly. These cracks start at the surface.

Cold Flow
Permanent deformation of a material due to mechanical force or pressure (not due to heat softening).

Cold Molding
A procedure in which a composition is shaped at room temperature and cured by subsequent baking.

Cold Plate
A heat transfer surface cooled by forced air or other heat transfer fluid to which heat-dissipating parts are mounted.

Cold Shuts
See also Incomplete fusion. An imperfect junction between two flows of metal in a mold; this is caused by the surface of the streams of metal chilling too rapidly, or in effect being chilled to the extent that fusion is impossible. This discontinuity may have the appearance of a crack or seam with smooth or rounded edges.

Cold Solder Joint
(1) A solder connection exhibiting poor wetting and a grayish porous appearance due to insufficient heat, inadequate cleaning prior to soldering, or excessive impurities in the solder solution. (2) A joint with incomplete bonding caused by insufficient application of heat to the base metal during soldering.

Cold Welding (CW)
A solid state welding process in which pressure is used at room temperature to produce coalescence of metals with substantial deformation at the weld.

Collimate
To render parallel.

Collimation
The process of aligning the optical axis of optical systems to the reference mechanical axes or surfaces of an instrument, or the adjustment of two or more optical axes with respect to each other. The process of making light rays parallel.

Collimator
An optical device that renders diverging or converging rays parallel. It may be used to simulate a distant target or to align the optical axes of instruments.

Combinational Logic Function
A logic function wherein, for each combination of states of the input or inputs, there corresponds one and only one state of the output or outputs. The terms *combinative* and *combinatorial* have also been used to mean *combinational*.

Combination of Adopted Items Drawing
A combination of adopted items drawing depicts the item constituting a combination of items and assigns a unique identification number to the combination. The drawing serves as the basic document for assignment of a stock number to the combination.

Command
An electronic pulse, signal, or set of signals to start, stop, change, or continue some operation.

Command and Control System
The facilities, equipment, communications, procedures, and personnel essential to a commander for planning, directing, and controlling operations of assigned forces pursuant to the missions assigned.

Command and Control System Equipment
The main mission element equipment and related ground equipment used in collecting, transmitting, processing, and displaying information for command and control.

Command Entry
A single command that causes a computer to perform a series of steps.

Command Language
A type of dialogue in which a user composes control entries with minimal prompting by the computer.

Commercial and Government Entity Code (CAGE Code)
A five-position alphanumeric code with a numeric in the first and last positions (e.g., 27340, 2A345, or 2AAA5), excluding the letters I and O assigned to U.S. organizations that manufacture and/or control the design of items supplied to a Government Military or Civil Agency

or assigned to U.S. organizations primarily for identifying contractors in the mechanical interchange of data required by MILSCAP and the Service/Agency automatic data processing (ADP) systems. Previously known as FSCM and Code Identification.

Commercial Drawing
Drawings prepared by a commercial design activity, in accordance with that activity's documentation standards and practices, to support the development and manufacture of a product not developed at Government expense.

Commercial Item
A product, material, component, subsystem, or system sold or traded to the general public in the course of normal business operations at prices based on established catalog or market prices.

Commercial Item Description (CID)
A simplified product description or specification that describes, by salient functional or performance characteristics, the available, acceptable commercial products that will satisfy the Government's needs.

Commercial Manuals
Manuals applicable to equipment designed and manufactured to commercial specifications, rather than military specifications, and used to support military equipment, systems, and facilities.

Commercial Off the Shelf (COTS) Item
A product, material, component, subsystem, or system sold or traded to the general public in the course of normal business operations at prices based on established catalog or market prices.

Commercial Off-the-Shelf Products
Products in regular production sold in substantial quantities to the general public or industry at established market or catalog prices.

Commercial Standard
A company document which establishes engineering and technical limitation and applications for items, materials, processes, methods, designs and engineering practices unique to that company.

Commodity Control List (CCL)
The list of items in the Export Administration Regulations at 15 CPR 399. Licenses from the Department of Commerce are required to export such items and the technical data relating to them.

Common Cause
A source of variation that affects all the individual values of the process output being studied. In control chart analysis, it appears as part of the random process variation.

Common Hand Tools
Items of tools found in common usage or applicable to a variety of operations or to a single operation on a variety of material. Screwdrivers, hammers, and wrenches are examples of common hand tools.

Common-Mode Interference
(1) Interference that appears between signal leads, or the terminals of a measuring circuit, and ground. (2) A form of coherent interference that affect two or more elements of a

network in a similar manner (i.e., highly coupled), as distinct from locally generated noise or interference that is statistically independent between pairs of network elements.

Common-Mode Rejection
The ability of a device to reject a signal that is common to both its input terminals.

Common-Mode Voltage
That amount of voltage common to both input terminals of a device.

Common Part
A part or component that is generic because (a) equivalent parts are available from more than one manufacturer and (b) it is not designed or intended for exclusive use in or by a single system or piece of equipment.

Common Return
A return path that is common to two or more circuits and that serves to return currents to their source or to ground.

Common Tool
A tool, routinely found in the tool supply of maintenance organizations for a similar class of system or equipment, signals is generic because it is available from more than one manufacturer and is not designed or intended for exclusive use on or with a single system or piece of equipment.

Company Standard
A company document that establishes engineering and technical limitations and applications for items, materials, processes, methods, designs, and engineering practices unique to that company.

Comparative Test
Comparative tests compare end item signal or characteristic values against a specified tolerance band and present the operator with a go or no-go readout; a "go" for signals within tolerances and a "no-go" for out-of-tolerance signals.

Comparator
(1) A device capable of comparing a measured value with predetermined limits to determine if the value is within these limits. (2) A device capable of comparing digital signals to determine agreement.

Comparison Tester
A device that uses a known good unit (golden unit) as a means for comparing test results with the unit under test when both are subjected to the same stimuli.

Compatible
The ability of different resin systems to be processed in contact with each other without degradation of end product properties.

Competent Manufacturer
A manufacturer capable of producing similar products at the same state of the art in the same or similar lines of technology.

Compile
(1) To translate a computer program expressed in a high-level language into a program expressed in an intermediate language, assembly language, or machine language. (2) To prepare a machine language program from a computer program written in another programming

language by making use of the overall logic structure of the program, or by generating more than one computer instruction for each symbolic statement, or both, as well as performing the function of an assembler.

Comiler
A computer program which translates high order language instructions of a computer program into machine language before the instructions are executed (for example, BASIC to machine code).

Compiler-Driven Simulation
The simulation carried out by translation of the network description into machine executable employed code. This technique is employed for clocked (synchronous) networks and primitive blocks.

Complete Operating Equipment
Complete operating equipment is defined as equipment, together with the necessary detail parts, accessories, and components, or any combination thereof, required for the performance of a specified operational function. Certain equipments may be complete within themselves and not require the addition of detail parts, accessories, or components to perform a specified operational function.

Complete Thread
Threads whose profile lies within the size limits.

Complement/Component Listing
Items or grouping of items, either nomenclature or non-nomenclatured, that comprise an item level and are essential for performing its intended functions are, issued atomically with the equipment and are considered "Part of" such equipment.

Complete Reference Designation
A reference designation that consists of a basic reference designation and, as prefixed, all the reference designations that apply to the subassemblies or assemblies within which the item is located, including those of the highest level needed to designate the item uniquely.

Complete Thread
The thread whose profile lies within the size limits.

Complex Feature
A single surface of compound curvature or a collection of other features that constrains up to six degrees of freedom.

Compliance
An affirmative indication or judgment that the supplier of a product or service has met the requirements of the relevant specifications, contract, or regulation; also, the state of meeting the requirements.

Compliant Printed Wiring Board
Compliant printed wiring boards meet all the requirements (including qualification) specified in the applicable specification sheet and printed board procurement documentation.

Component
(1) A functional unit that is viewed as an entity for purposes of analysis, manufacturing, maintenance, or record keeping. Examples are hydraulic actuators, valves, batteries, electrical harnesses, and individual electronic boxes. (2) An assembly, or part thereof, that is essential to the operation of some larger assembly. It is an immediate subdivision of the assembly to

which it belongs. *Note:* the proper usage of the term is dependent on the frame of reference. A radio receiver may be considered to be a component of a complete radio set (transmitter-receiver) if the radio set is part of a larger system. The same receiver could also be considered as a subsystem, in which case the IF amplifier section would be a component of the receiver but not of the radio set. Similarly, a resistor, capacitor, vacuum tube, transistor, or other item within the IF amplifier section is a component of that section. (3) A part or combination of parts, having a specified function, which can only be installed or replaced as a whole and is also generally expendable.

Component Density
The quantity of components on a printed board per unit area.

Component Hole
A hole used for the attachment and electrical connection of component terminations, including pins and wires, to the printed board.

Component Lead
A solid or stranded wire or formed conductor that extends from a component and serves as a mechanical or electrical connection, or both.

Component Orientation
Component orientation is the direction in which the components, on a printed board or other assembly, are lined up physically with respect to the polarity of polarized components, and also with respect to one another and to the board.

Component Side
That side of the printed board on which most of the components will be mounted.

Composite
Composites are combinations of materials differing in composition or form on a macro scale. The constituents retain their identities in the composite. Normally, the components can be physically identified, and there is an interface between them.

Composite Armor
Composite armor is an armor configuration consisting of two or more different armor materials bonded together to form a protective unit.

Composite Class
A major subdivision of composite construction in which the class is defined by the fiber system and the matrix class, e.g., organic-matrix filamentary laminate.

Composited Circuit
A circuit that can be used simultaneously for telephony and DC telegraphy, or signaling; separation between the two being accomplished by frequency discrimination.

Composite Material
Composites are considered to be combinations of materials differing in composition or form on a macro scale. The constituents retain their identities in the composite; that is, they do not dissolve or otherwise merge completely into each other, although they act in concert. Normally, the components can be physically identified and exhibit an interface between one another.

Composition
This term is used in classifying commodities that are differentiated strictly by their respective chemical composition. It is designated in accordance with accepted trade practice.

Compound Mount
A three-element holding or support device consisting of an intermediate mass contained between resilient elements used for protection of a supported device from vibrations or acceleration. Also called a *two-stage mount*.

Comprehensive Testability
An overall testability design characteristic that includes both hardware design and test design.

Compression
An operation performed on raster image data to remove redundant information and thus reduce the stored or interchanged size. Negative compression is the case when this operation results in an increase rather than a decrease in size.

Compression Molding
A technique of thermoset molding in which the molding compound (generally preheated) is placed in the open mold cavity, the mold is closed, and heat and pressure (in the form of a downward moving ram) are applied until the material is cured.

Compressive Strength
Pressure load at failure of a shaped specimen divided by a cross-sectional area of the specimen, which is usually the original sectional area.

Compromise
(1) The known or suspected exposure of clandestine personnel, installations, or other assets or of classified information or material, to an unauthorized person. (2) The disclosure of cryptographic information, or recovery of plain text of encrypted messages by unauthorized persons through cryptanalysis methods. (3) The disclosure of information or data to persons not authorized to receive it. (4) A violation of the security policy of an automated information system such that an unauthorized disclosure of sensitive information may have occurred.

Compromising Emanations
Unintentional intelligence-bearing signals that, if intercepted and analyzed, disclose the national security information transmitted, received, handled, or otherwise processed by any classified information processing system.

Computer Aided Design (CAD)
A process that uses a computer system to assist in the creation, modification, and display of a design.

Computer Data
A representation of facts, concepts, or instructions in a structured communication among computer equipment. Such data can be external (in computer-readable form) or resident within the computer equipment and can be in the form of analog or digital signals.

Computer Database
A collection of data in a form capable of being processed and operated by a computer.

Computer Data Definition
A statement of the characteristics of the basic elements of information operated upon by hardware in responding to computer instructions. These characteristics may include, but are not limited to, type, range, structure, and value.

Computer Firmware
An assembly composed of a hardware unit and a computer program integration to form a functional entity whose configuration cannot be easily altered during normal operation.

Computer Graphics
(1) Methods and techniques for converting data to or from graphic displays via computers. (2) That branch of science and technology that is concerned with methods and techniques for converting data to or from visual presentation using computers.

Computer Graphics Metafile (CGM)
Standard for the description, storage, and communication of graphical information in a device-independent manner.

Computer Hardware
Devices capable of accepting and storing computer data, executing a systematic sequence of operations on computer data, or producing control outputs. Such devices can perform substantial interpretation, computation, communication, control, or other logical functions.

Computer Language
A language that is used to program a computer. The language may be a high-level language, an assembly language, or a machine language.

Computer Network
(1) A network of data processing nodes that are interconnected for the purpose of data communication. (2) A complex consisting of two or more interconnected computers.

Computer-Oriented Language
A programming language whose words and syntax are designed for use on a specific computer or class of computers.

Computer Program/Software
A series of instructions that direct a computer to perform a sequence of operations that produce a desired output. This program may be stored on one or more physical media such as optical disk, magnetic tape, magnetic disks, or punched cards. The terms *computer program* and *computer software* are used synonymously.

Computer Resources
The totality of computer hardware, software, personnel, documentation, supplies, and services applied to a given effort.

Computer Science
The branch of science and technology that is concerned with methods and techniques relating to data processing performed by automatic means.

Computer Software (or Software)
A combination of associated computer instructions and computer data definitions required to enable the computer hardware to perform computational or control functions. Software also includes microcode and firmware. Components of nuclear safety critical software are categorized by the degree of involvement in nuclear critical functions:

Category I software (direct involvement). That software which is directly involved with the display of data of the control, monitoring, or execution of nuclear critical functions.
Category II software (indirect involvement). That software which directly or indirectly transfers, stores, or shares data with or control Category I software. Category II software is divided into subcategories:

Subcategory II A software: System run-time support software.

Subcategory II B software: Applications-oriented software. This includes not only tactical software but also application-unique support software.

Category III software (no involvement). That software which is neither Category I nor Category II.

Computer Software Component

(1) A distinct part of a computer software configuration item (CSCI). Computer software components may be further decomposed into other computer software components and computer software units. (2) A computer software component is a functionally or logically distinct part of a computer software configuration item that is distinguished for purposes of convenience in designing and specifying a complex computer software configuration item as an assembly of subordinate elements.

Computer Software Configuration Item (CSCI)

A computer software configuration item is a configuration item (CI) for computer software.

Computer Software Documentation

Owner's manuals, user's manuals, installation instructions, operating instructions, and other similar items, regardless of storage medium, that explain the capabilities of the computer software or provide instructions for using the software. *Note:* the term *software life cycle data* is also used to address software documentation.

Computer Software Unit

An element specified in the design of a computer software component that is separately testable.

Concave

A term denoting a hollow, curved surface.

Concentricity

Concentricity is the condition in which the axes of all cross-sectional elements of a surface of revolution are common to the axis of a datum feature.

Concept Exploration Phase

(1) The identification and exploration of alternative solutions or solution concepts to satisfy a validated need. (2) That part of the acquisition life-cycle when alternative concepts are explored and evaluated. (3) The first phase in the material life cycle. The phase in which the technical, military, and economic basis for the program, and concept feasibility, are established through pertinent studies.

Conceptual Design Data

Data, such as drawings or 3D solid models, that describe the engineering concepts on which a proposed technology or design approach is based.

Conceptual Design Drawings

Drawings that describe the engineering concepts on which a proposed technology or design approach is based. They are used when there is a need to verify preliminary design and engineering and confirm that the technology is feasible and that the design concept has the potential to be useful in meeting a specific requirement.

Conditioning

Time-limited exposure of a test specimen to a specified environment prior to testing.

Conducted Interference
Undesired signals that enter or leave equipment along a conductive path.

Conductive Foil
A thin sheet of metal that may cover one or both sides of the base material and is intended for forming the conductive pattern.

Conductive Pattern
The configuration or design of the conductive material on the base material. (Includes conductors, lands, and through connections when these connections are an integral part of the manufacturing process.)

Conductor
(1) A single conductive path in a conductive pattern. (2) A wire or combination of wires not insulated from one another, suitable for carrying current.

Conductor Base Width
The conductor width at the plane of the surface of the base material.

Conductor Layer
The total conductive patterns formed upon one side of a single layer of base material.

Conductor Pattern
- see CONDUCTIVE PATTERN.

Conductor Spacing
The distance between adjacent edges (not centerline to centerline) of isolated conductive patterns in a conductor layer.

Conductor Thickness
The thickness of the conductor including all metallic coatings. (It excludes nonconductive protective coatings.)

Conductor Width
The observable width of a conductor at any point chosen at random on the printed board, normally viewed vertically from above unless otherwise specified. (Imperfections for example, nicks, pinholes, or scratches allowable by the relevant specifications shall be ignored.)

Cone of Tolerance
The specification of tighter test tolerances at the factory (or at component or subassembly indenture levels) that gradually loosen at successive maintenance levels (or higher assembly indenture levels, system level being the highest). Also known as *inverted pyramid*. The use of a cone of tolerance approach tends to reduce retest OK problems.

Confidence Test
(1) A test performed to provide a high degree of certainty that the unit under test is operating acceptably. (2) A check of the performance of all test system stimulus and measurement functions to detect degradation with respect to system specifications and to inform the system operator.

Configuration
(1) The functional and physical characteristics of existing or planned hardware, firmware, software, or a combination thereof as set forth in technical documentation and ultimately achieved in a product. (2) The collection of interconnected objects that make up a system or

subsystem. (3) The total software modules in a software system or hardware devices in a hardware system and their interrelationships.

Configuration Audit
 - see FUNCTIONAL CONFIGURATION AUDIT and PHYSICAL CONFIGURATION AUDIT.

Configuration Baseline
Configuration documentation formally designated by the Government at a specific time during a CI's life cycle. Configuration baselines, plus approved changes from those baselines, constitute the current approved configuration documentation. There are three formally designated configuration baselines in the life cycle of a configuration item, namely the functional, allocated, and product baselines.

Configuration Control
(1) The systematic proposal, justification, evaluation, coordination, approval, or disapproval of proposed changes, and the implementation of all approved changes, in the configuration of a CI after establishment of the configuration baseline for the CI. (2) After establishing a configuration, such as that of a telecommunications or computer system, the evaluating and approving changes to the configuration and to the interrelationships among system components. (3) In distributed-queue dual-bus networks (DQDB), the function that ensures the resources of all nodes of a DQDB network are configured into a correct dual-bus topology. *Note:* the functions that are managed include the head of bus, external timing source, and default slot generator functions.

Configuration Control Board (CCB)
A board composed of technical and administrative representatives who recommend approval or disapproval of proposed engineering changes to a CI's current approved configuration documentation. The board also recommends approval or disapproval of proposed waivers and deviations from a CCI's current approved configuration documentation.

Configuration Documentation
The technical documentation that identifies and defines the item's functional and physical characteristics. The configuration documentation is developed, approved, and maintained through three distinct evolutionary increasing levels of detail. The three levels of configuration documentation are the functional configuration documentation, the allocated configuration documentation, and the product configuration documentation.

Configuration Identification
(1) The selection of the documents to compose the baseline for the systems and configuration items involved, and the numbers and other identifiers affixed to the items and documents. The approved documents that identify and define the item's functional and physical characteristics in the form of specifications, drawings, associated lists, interface control documents, and drawings referenced therein. The configuration identification is developed and maintained through three distinct evolutionary increasing levels of detail, each used for establishing a specific baseline. (2) Configuration identification includes the selection of CIs; the determination of the types of configuration documentation required for each CI; the issuance of numbers and other identifiers affixed to the CIs and to the technical documentation that defines the CIs' configuration, including internal and external interfaces; the release of CIs and their associated configurations documentation; and the establishment of configuration baselines for CIs.

Configuration Item (CI)
An aggregation of hardware, firmware, software, or any of its discrete portions, that satisfies an end-use function and is designated for configuration management. CIs may vary widely in

complexity, size, and type, and from an aircraft, ship, or electronic system to a test meter or round of ammunition. During development and manufacturer of the initial (prototype) production configuration, CIs are those items whose performance parameters and physical characteristics must be separately defined (specified) and controlled to provide management insight needed to achieve the overall end-use function and performance. Any item required for logistic support and designated for separate procurement is a CI.

Configuration Item Identification (CII Number)
The alphanumeric number assigned to identify a configuration item. When assigned, it is the unchanging base number to which serial numbers are assigned.

Configuration Management Plan (CMP)
The document defining how configuration management will be implemented (including policies and procedures) for a particular acquisition or program.

Conformal Coating
An insulating protective coating that conforms to the configuration of the object coated, applied to the completed board assembly.

Conformance Tests
(1) Tests that are specifically made to demonstrate conformity with applicable standards or specifications. (2) The testing of a candidate product for the existence of specific characteristics required by a standard; testing the extent to which an item under test is a conforming implementation.

Connection Diagram (Wiring Diagram)
A diagram that shows the connections of an installation or its component devices or parts. It may cover internal or external connections, or both, and contains such detail as is needed to make or trace connections that are involved. The Connection Diagram usually shows general physical arrangement of the component devices or parts.

Connector
Any device used to provide rapid connect/disconnect service for electrical cable and wire terminations.

Consignee (Receiver)
Party to whom materiel is shipped and whose name and address appear in the "ULTIMATE CONSIGNEE OR MARK FOR" block of the shipping label.

Consignor (Shipper)
Party who ships materiel and whose name and address appear in the "FROM" block of the shipping label.

Consolidation Container
A container used to consolidate more than one line item into a single shipping container to be shipped to one destination, but not necessarily to one addressee.

Constraint
A limit to one or more degrees of freedom.

Contact Area
The common area between a conductor and a connector through which the flow of electricity takes place.

Contact Resistance
The electrical resistance of the metallic surfaces at their interface in the contact area under specified conditions.

Contact Spacing
The distance between the centerlines of adjacent contact areas.

Continuity
An uninterrupted path for the flow of electrical current in a circuit.

Continuity Test
A test for the purpose of detecting broken or open connections and ground circuits in a network or device.

Continuous Feature (CF)
A group of two or more interrupted features as a single feature. It is used to identify a group of two or more features of size where there is a requirement that they be treated geometrically as a single feature of size.

Continuously Degrading Faults
Faults that permit continued use of the equipment for a limited time. However, if operation continues for protracted times, the system will transition into a catastrophic fault.

Continuous Weld
A weld that extends continuously from one end of a joint to the other. Where the joint is essentially circular, it extends completely around the joint.

Contour Definition Drawing
A contour definition drawing contains the mathematical, numeric, or graphic definition required to locate and define a contoured surface. It does not establish item identification for the items delineated thereon. The contour of a part is defined on a detail drawing or by reference to a contour definition drawing.

Contract
A mutually binding legal relationship obligating the seller to furnish the supplies or services (including construction) and the buyer to pay for them. It includes all types of commitments that obligate the Government to an expenditure of appropriated funds and that, except as otherwise authorized, are in writing. In addition to bilateral instruments, contracts include, but are not limited to, awards and notices of awards; job orders or task letter issued under basic ordering agreements; letter contracts; orders, such as purchase orders, under which the contract becomes effective by written acceptance or performance; and bilateral contract modifications.

Contract Data Requirements List (CDRL)
A contract form that provides, in one place in the contract, a list of data items required to be delivered under the contract, with the exception of data specifically required by standard clauses of the Federal/Defense acquisition regulations.

Contract Delivery
Some form of transfer of contracted product or performance of a contracted service. Contract delivery may involve some physical transfer or relocation of the data product. However, leaving data in place and tagging it in such a way as to logically transfer designated data may also be considered delivery.

Contracting Activity
That Government activity having a legal agreement or order with an individual, partnership, company, corporation, association, or other entity for the design, development, manufacture, maintenance, modification, or supply of items or services.

Contracting Officer
A contracting officer is a person with the authority to enter into, administer, or terminate contracts and make related determinations and findings. The term includes authorized representatives of the contracting officer acting within the limits of their authority as delegated by the contracting officer.

Contractor
An individual, partnership, company, corporation, association, or other service having a contract with the procuring activity for the design, development, manufacture, maintenance, modification, or supply of items under the terms of a contract. A Government activity performing any or all of the above functions is considered to be a contractor for engineering drawing preparation purposes.

Contractor, Associate
Any contractor subordinate to the prime contractor. Included under this heading are subcontractors, Group B contractors, vendors, and suppliers.

Contractor Enterprise
The prime contractor, teamed contractors, subcontractors, suppliers, and vendors working together.

Contractor, Prime
The contractor with responsibility for designing, integrating, and producing the overall system. Included under this heading are integrating contractors, airframe contractors, and Group A contractors.

Contract Quality Requirements
(1) The detailed requisites for quality incumbent on the contactor, consisting of (1) all quality requirements contained in a contract, and (2) the detailed contractual requisites provided by the contract incumbent on the contractor to substantiate conformance of product or service to quality requirements of the contract.

Control
A major subdivision within configuration management. The procedures by which changes to the design requirements are proposed and formally processed.

Control Data
Data that select an operating mode or sub-mode in a program, direct the sequential flow, or otherwise directly influence the function of a program.

Control Designation Symbol
A symbol that identifies the particular manner, permissible or required, in which an input variable (possibly in combination with other variables) causes the logic element to perform according to its defined function.

Control Entry
User input of data for computer processing/sequence control (e.g., function key actuation, menu selection, and command entry) whereby the character or symbol and control key are jointly activated, requiring computer responses to such inputs.

Controller
In an automated radio, the device that commands the radio transmitter and receiver, and that performs processes, such as automatic link establishment, channel scanning and selection, link quality analysis, polling sounding, message store and forward, address protection, and anti-spoofing.

Control Limits
Control limits are criteria that establish the maximum variation beyond which action must be taken to investigate, and when feasible, correct the cause(s) of nonconformance. Control limits do not preclude corrective action when abnormal patterns of variation occur without any individual data exceeding the control limits. Control limits are developed using standard statistical methods or other approved techniques and are based on documented process history. They are established to assist in fulfilling the contractor's responsibility for submitting a conforming item, identifying necessary corrective actions, and reducing nonconformance levels.

Control Logic
The topological connectivity and the set of conditions that together govern the apparent sequencing of operations within a process (or among concurrent processes).

Control Vane
A moveable tab or aerodynamic surface attached to a missile airframe and used to control the flight attitude of the vehicle.

Conversational Mode
A mode of operation of a data processing system in which the user of a terminal carries on a dialogue with the computer such that each unit of input entered by the user elicits a prompt response from the computer. Synonymous with *interactive mode*.

Convex
A term denoting a surface resembling the outside of a sphere or ball.

Cookie
Is simply small bits of data that is commonly transmitted from a web server to a web browser. Cookies can also be entirely processed client-side. The browser stores the message in a text file, and each time the browser requests from the server a particular page, the message is sent back to the server. Cookies can be accessed, read, and used by malicious websites unintentionally visited by innocent users. This cookie information can be used to gather intelligence on the user and later used against the user, or the cookie information can be used to access the original website.

Cooling, Regenerative
The cooling of a part of an engine by the fuel or propellant being delivered to the combustion chamber; specifically, the cooling of a rocket engine combustion chamber or nozzle by circulating the fuel or oxidizer, or both, around the part to be cooled.

Coordinated Military Specification
A coordinated military specification is a document required by more than one military department and is coordinated with various activities of the interested departments.

Copolymer
A compound resulting from the chemical reaction of two chemically different monomers with each other.

Copy Card
An unprocessed tabulating card whose aperture contains undeveloped sensitized diazo microfilm, not camera microfilm.

Cord Circuit

A switchboard circuit terminated in two plug-ended cords, used to establish connections manually between user lines or between trunks and user lines. A number of cord circuits are furnished as part of the manual switchboard position equipment. The cords may be referred to as *front cord* and *rear cord* or *trunk cord* and *station cord*.

Corner Mark (Crop Mark)

A mark at the corners of a printed board artwork, the inside edges of which usually locate the borders and establish the contour of the board.

Corner Radius

The convex radius on the surface of a part connecting intersecting surfaces.

Corrective Action

Changes to process work instructions, workmanship practices, training, inspections, tests, procedures, specifications, drawings tools, equipment facilities, resources, or materials that result in preventing, minimizing, or eliminating non-conformances.

Corrective Action Board (CAB)

A contractor board consisting of management representatives of appropriate contractor organizations with the level of responsibility and authority necessary to ensure the prevention of nonconformances, to manage quality improvement efforts as appropriate, to access and manage nonconformance cost elimination, to ensure that causes of nonconformances are identified, and to ensure that corrective actions are effected throughout the contractor's organization.

Corrective Maintenance

(1) The maintenance carried out after a failure has occurred and intended to restore an item to a state in which it can perform its required function. (2) Actions performed to restore failed or degraded equipment. It includes fault isolation, repair or replacement of defective circuit cards, or components alignment and check-out.

Corrosion

A specific type of deterioration resulting in damage or impairment of metals or metallic parts as the result of attack by moisture, air, acid, alkali, chemicals, or electrochemical action.

Corrosion-Erosion

An accelerated corrosion process caused by the abrasive action of a moving liquid, especially one that contains suspended particles.

Corrosion Fatigue

A failure of a metal resulting from the combined action of corrosion and fatigue (cyclic stressing).

Corrosion Potential

A potential of a corroding surface in an electrolyte relative to a reference electrode (also known as *rest potential, open-circuit potential,* or *freely-corroding potential*).

Corrosion Rate

A measure of the speed (usually averaged) at which corrosion proceeds, expressed as a weight loss per unit area or thickness change per unit time. Alternatively, the corrosion current density (in amps per unit area) gives a measure of the instantaneous rate of corrosion for specified conditions of the environment at a given time.

Corrosion Resistant Steels
Steel with sufficient alloy content (usually chromium and nickel) to resist atmospheric corrosion; sometimes called "stainless" steels, although they are neither stainless nor rustless. A wide variety of analyses and properties are included in the term *corrosion-resistant steels.*

Cosmetic Solid Model
A computer aided design (CAD) 3D dimensional solid model sufficiently defined to provide a visual understanding of the item, but not fully defined in all respects.

Countermeasures
The art of employing devices and techniques to impair the operational effectiveness of enemy activity, such as anti-missile defense systems.

Countermeasures, Active Electronic
Countermeasure devices or techniques, based on the use of electronic systems, of such a nature that their use can be detected by the enemy.

Countermeasures, Chaff
Narrow metallic strips that reflect electromagnetic radiation and can be dispensed by incoming missile or aircraft to create radar echoes that confuse enemy radars.

Countermeasures, Electronic
The use of electronics to reduce the effectiveness of enemy equipment or tactics that can be affected by electromagnetic radiation.

Countermeasures, Passive Electronic
Electronic countermeasures that are of such a nature that their use cannot be detected by the enemy.

Counterpoise
A system of wires or other conductors elevated above and insulated from the ground, forming the lower system of conductors of an antenna. For overhead lines, a conductor or system of conductors arranged beneath the line, located on, above, or most frequently below the surface of the earth, and connected to the footings of the towers or poles supporting the line.

Countersink
A bevel or flare at the end of a hole.

Coupling
Energy transfer between circuits, equipments, or systems.

Coupling Agent
(1) That part of a sizing or finish that is designed to provide a bonding line between the reinforcement and the laminating resin. (2) Any chemical substance designed to react with both the reinforcement and matrix phases of a composite material to form or promote a stronger bond at the interface. Coupling agents are applied to the reinforcement phase from an aqueous or organic solution or from a gas phase, or added to the matrix as an integral blend.

Coupling Coefficient
A measure of the electrical coupling that exists between two circuits, equal to the ratio of the mutual impedance to the square root of the product of the self-impedances of the coupled circuits, all impedances being of the same kind.

Coupling, Conducted
Energy transfer through a conductor.

Coupling, Free-Space
Energy transfer via electromagnetic fields not in a conductor.

Coupon
 - see TEST COUPON.

Cracking
A condition consisting of breaks in metallic or nonmetallic coatings, or both, that extend through to an underlying surface.

Crazing (Conformal Coating)
(1) A network of fine cracks on the surface or within the conformal coating. (2) Fine resin cracks at or under the surface of a plastic.

Creep
The dimensional change with time of a material under load that follows the initial instantaneous elastic deformation. Creep at room temperature is called *cold flow.*

Crest
The surface of a thread that joins the flanks of the thread and is farthest from the cylinder or cone from which the thread projects.

Crest Diameter
The diameter of an imaginary cylinder or cone bounding the crest of a screw thread. This is the major diameter of an external thread or the minor diameter of an internal thread.

Crest Truncation
The crest truncation of a thread is the radial distance between the sharp crest (crest apex) and the cylinder or cone that would bound the crest.

Crest Width
The distance between the points of intersection of the flanks of the thread ridge and the imaginary cylinder defined by the crest diameter.

Crevice Corrosion
A localized corrosive attack resulting from the formation of a concentration cell in a crevice between two metal surfaces or a metal and nonmetal surface.

Critical Application Item
Any item essential to preserving human life or that, if it fails, endangers human life or adversely affects the completion of a military operation directly or through the impact of its failure on an end item or system.

Critical Confusion
When a safety symbol elicits the opposite or prohibited action. For instance, when a safety symbol meaning "no fires allowed" is misunderstood to mean "fires allowed here."

Critical Defect
A critical defect is a defect that judgment and experience indicate would result in hazardous or unsafe conditions for individuals using, maintaining, or depending upon the product, or a defect that judgment and experience indicate is likely to prevent performance of the tactical function of a major end item such as a ship, aircraft, tank, missile, or space vehicle.

Critical Defective
A critical defective is a unit or product that contains one or more critical defects and may also contain major and/or minor defects.

Critical Design Review (CDR)
This review shall be conducted for each configuration item when detail design is essentially complete. The purpose of this review will be to (a) determine that the detail design of the configuration item under review satisfies the performance and engineering specialty requirements of the configuration item development specifications, (b) establish the detail design compatibility among the configuration item and other items of equipment, facilities, computer programs and personnel, (c) assess configuration item risk areas (on a technical), cost, and schedule basis, (d) assess the results of the producibility analysis conducted on system hardware, and (e) review the preliminary product specifications.

Critical Failure
A failure or combination of failures that prevent an item from performing a specified mission or put human life at risk.

Critical Flaw Size
A size of flaw in a structure that will cause failure of the structure at the expected operational stress level.

Critical Manufacturing Process
A process that is mandatory for use during the manufacturing of an item and without which an acceptable item cannot be produced.

Critical Nonconformance
A nonconformance that judgment and experience indicate is likely to result in hazardous or unsafe conditions for individuals using, maintaining, or depending upon the supplies or services, or is likely to prevent performance of a vital agency mission.

Critical Safety Characteristic
Any feature (such as tolerance, finish, material composition, manufacturing, assembly or inspection process, or product) that if nonconforming or missing could cause the failure or malfunction of the critical safety item.

Critical GSE (Ground Support Equipment)
GSE whose loss of function or improper performance could result in serious personnel injury, damage to flight hardware, loss of mission, or major damage to a significant ground asset.

Critical Safety Item (CSI)
A part, assembly, installation, or production system with one or more critical characteristics that, if not conforming to the design data or quality requirements, would result in an unsafe condition. Unsafe conditions relate to hazard severity categories of System Safety Program Requirements and include conditions that could cause loss or serious damage to the end item or major components, loss of control, or serious injury to personnel.

Critical Technology
Technologies that consist of (a) arrays of design and manufacturing know-how (including technical data), (b) keystone manufacturing, inspection, and test equipment, (c) keystone materials, (d) goods accompanied by sophisticated operation, application, or maintenance know-how that would make a significant contribution to the military potential of any country or combination of countries and that may prove detrimental to the security of the U.S. Also referred to as *militarily critical technology.*

Crop Mark
- see CORNER MARK.

Cross-Connection
Connections between terminal blocks on the two sides of a distribution frame or between terminals on a terminal block. *Note:* connections between terminals on the same block are also called *straps*.

Cross Coupling
Undesired signal coupling between two or more different communication channels, circuit components, or parts.

Cross Hatching
The breaking of large conductive areas by the use of a pattern of the desired signal by an undesired signal.

Cross-Linking
The setting up of chemical links between the molecular chains.

Cross-Ply
Any filamentary laminate that is not uniaxial. Same as *angle-ply*. In some references, the term cross-ply is used to designate only those laminates in which the lamina are at right angles to one another, while the term *angle-ply* is used for all others.

Crosstalk
(1) The undesirable interface caused by the coupling of energy between signal paths. (2) Signal interference between nearby conductors caused by magnetic field effects.

Cruciform Configuration
An aerodynamic configuration consisting of identical control and stabilization surfaces that are located at right angles to each other around the body of a missile airframe.

Crystal, Piezoelectric
A natural substance, such as quartz or tourmaline, capable of producing a voltage when under stress or pressure, or producing pressure when under an applied voltage. Under stress, it has the property of responding only to a given frequency when cut to a given thickness. It is therefore valuable for transmitters and oscillators whose frequencies range between 500 kHz and 10 MHz.

C-Stage
The final stage of the curing reaction of a thermosetting resin in which the material has become practically infusible and insoluble.

Cube-Root Law
Cube-root law is a scaling law applicable to many blast phenomena. It relates the time and distance at which a given blast effect is observed to the cube root of the energy yield of the explosion.

Cumulative Form Variation
The combined effect on functional size of individual thread variations in lead (pitch), helix, flank angle, taper, and roundness. It is the maximum difference between GO functional diameter size and pitch diameter size taken along and around the axis of the usable thread.

Cumulative Pitch
The distance measured parallel to the axis of the thread between corresponding points on any two threads regardless of whether they are in the same axial plane.

Cure
To irreversibly change the properties of a thermosetting resin by chemical reaction, e.g., condensation, ring closure, or addition. Cure may be accomplished by addition of curing (cross-linking) agents, with or without heat.

Cure Cycle
The schedule of time periods at specified conditions to which a reacting thermosetting material is subjected in order to reach a specified property level.

Cure Stress
A residual internal stress produced during the curing cycle of composite structures. Normally, these stresses originate when different components of a lay-up have different thermal coefficients of expansion.

Current
The rate of transfer of electricity. Practical unit is the ampere, which represents the transfer of one coulomb per second.

Current Absorption
Current proportional to the rate of accumulation of electric charges within an anistropic dielectric. The rate of accumulation, and hence the absorption current, decreases with time after any change of the potential gradient, and occurs with both an increase and a decrease of potential gradient so that the absorption current is reversible.

Current-Carrying Capacity
The maximum current that can be carried continuously, under specified conditions, by a conductor without causing objectionable degradation of electrical or mechanical properties of the printed board.

Current Density
A current expressed as a function of the area of the metal through which it flows: expressed as amps per square meter, milliamps per square decimeter, and so on.

Current Design Activity (CDA)
An activity (Government or contractor) currently having responsibility for the design of an item and the preparation or maintenance of drawings and associated documents. Current design activity could be the original activity or new activity when that responsibility is transferred from another Government or contractor design activity.

Cutoff Frequency
The frequency below which electromagnetic energy will not propagate in a waveguide.

Cyclic Distortion
In telegraphy, distortion that is neither characteristic, bias, nor fortuitous, and which in general has a periodic character. *Note:* its causes are, for example, irregularities in the duration of contact time of the brushes of a transmitter, distributor, or interference by disturbing alternating currents.

Cylindricity
The condition of a surface of revolution in which all points of the surface are equidistant from a common axis.

D

Daisy Chain
In electrical and electronic engineering, a daisy chain is a wiring scheme in which device A is wired to device B, device B is wired to device C, device C is wired to device D, and so forth. Connections do not form webs (device C cannot be directly connected to device A), nor do they loop back from the last device to the first. Daisy chains may be used for power, analog signals, digital data, or a combination thereof.

Damage
The partial or total loss of hardware caused by component failure; exposure of hardware to heat, fire, or other environments; human errors or other inadvertent events or conditions.

Damping
(1) The progressive diminution with time of certain quantities characteristic of a phenomenon. (2) The progressive decay with time in the amplitude of the free oscillations in a circuit.

Dart Configuration
A missile airframe that has the surfaces used for trim and control attached to the rear portion of the body, and the main lifting surfaces attached to the forward portion.

Data
(1) Recorded information, regardless of medium or characteristics, of any nature, including administrative, managerial, financial, and technical. (2) Data is a representation of facts, concepts, or instructions in a structured form suitable for acceptance, interpretation, or processing by computer equipment. Data can be external (in computer-readable form) or resident within the computer equipment and can be in the form of analog or digital signals.

Data Analysis
Indexing term. Refers to the application of statistical procedures to raw data to obtain information relating to some aspect of software development, use, reliability, or maintenance.

Data Area Titles
Data areas composed of information in machine-readable or human-readable form. *Note*: data areas are identified with the corresponding data area title in human-readable text that may be prefixed, if relevant, by the appropriate identifier.

Data Bank
A comprehensive collection of libraries of data, often a flat file but usually containing only data to support one application rather than the enterprise. Contrast with *database*.

Database
A collection of related data stored in one or more computerized files in a manner that can be accessed by users or computer programs via a database management system.

Database Partition
A logical or physical separation of data elements that facilitates change control, security, auditability, data usage control, computer resource control, and/or cost accounting control as required by contract.

R. Hanifan, *Concise Dictionary of Engineering: A Guide to the Language of Engineering*, DOI 10.1007/978-3-319-07839-7_4, © Springer International Publishing Switzerland 2014

Data Burst
Burst that contains data packets (such as point-to point, broadcast, and multicast bursts).

Data Carrier
A physical pattern or structure that contains encoded machine-readable characters. The carrier can be a structured pattern of markings, such as an ID or 2D symbol.

Data Depository
A facility designated to act as custodian in establishing and maintaining a master engineering specification and drawing depository service for Government-approved documents that are the property of the U.S. Government.

Data Dictionary
A repository of information describing the characteristics of data elements used in information systems and databases.

Data Display
(1) Output of data from a computer to its user. Generally, the phrase denotes visual output, but it may be qualified to indicate a different modality, such as an "auditory display." (2) Communication of visual, audio, or other output from a computer to its users.

Data Element
A basic unit of information representing characteristics (attributes) of a data item instance.

Data Entry
User input of data for computer processing and computer responses to such inputs.

Data Field
An area of the display screen reserved for user entry of a data item.

Data Identifier (DI)
A specified character (or string of characters) that defines the general category or intended use of the data that follows. *Note*: ASC MH10 Data Identifiers have a format of one alphabetic character alone, or one alphabetic character prefixed by one, two, or three numeric characters.

Data Item
(1) A set of characters of fixed or variable lengths that forms a single unit of data. (2) Deliverable data specified by the Contract Data Requirements List (CDRL) in which format and content is defined by the Data Item Description (DID).

Data Item Description (DID)
A standardization document that defines the data content, preparation instructions, format, and intended use of data required of a contractor.

Data Item Instance
A discrete part of a data item such as a single drawing from the data item set of engineering drawings and associated lists or a specific version of a technical report.

Data Link
The means of connecting one location to another for the purpose of transmitting and receiving data.

Data List (DL)
A tabulation of all engineering drawings, associated lists, specifications, standards, and subordinate data list pertaining to the item to which the data list applies and essential in-house documents necessary to meet the technical design disclosure requirements, except for those in-house documents referenced parenthetically.

Data Processing
The systematic performance of operations upon data such as handling, merging, sorting, and computing. *Note*: the semantic content of the data may or may not be changed.

Data Processing System
A system used to collect, process, and reproduce data in a selected format through the use of electronic or other automated equipment.

Data Protection
Functional capabilities that guard against unauthorized data access and tampering, user errors, and computer failure.

Data Qualifier
A specified character (or string of characters) that immediately precedes a data field that defines the general category or intended use of the data that follows.

Data Rights
The rights to use, disclose, reproduce, prepare derivative works, distribute copies to the public, and perform publicly, data in any manner and for any purpose, and to have or permit others to do so. Data shall mean recorded information, regardless of form or the media on which the information may be recorded, including technical data and computer software. Data does not include information incidental to contract administration, such as financial, administrative, cost or pricing, or management information.

Data Standard
A specific set of data entities, relationships among data entities, and their attributes, often expressed in the form of a Data Dictionary and a set of rules that govern data definition, data integrity, and data consistency.

Data Terminal
A device that is the source or destination of the user data processed by the data controller.

Data Transmission
Message exchange among system users, and also message exchange with other systems. Transmitted data may include numbers, words, pictures, or other representations of data.

Data Universal Numbering System (D U N S)
A nine-digit number assigned by Dun & Bradstreet to each business location in its global database; widely used as a tool for identifying, organizing, and consolidating information about businesses.

Data Validation
Functional capabilities that check data entry items for correct content or format as defined by software.

Datum
(1) A theoretically exact point, axis, or plane derived from the true geometric counterpart of a specified datum feature. A datum is the origin from which the location or geometric

characteristics of features of a part are established. (2) A theoretically exact point, axis, line, plane, or combination thereof derived from the theoretical datum feature simulator.

Datum Axis
The axis of a datum feature simulator established from the datum feature.

Datum Center Plane
The center plane of a datum feature simulator established from the datum feature.

Datum Feature
(1) An actual feature of a part that is used to establish a datum. (2) A feature that is identified with either a datum feature symbol or a datum target symbol.

Datum, Simulated
A point, axis, line, or plane (or combination thereof) coincident with or derived from processing or inspection equipment such as the following simulators: a surface plate, a gage surface, a mandrel, or mathematical simulation.

Datum Feature Simulator
Encompasses two types: Theoretical and physical.

Datum Feature Simulator (Theoretical)
The theoretically perfect boundary used to establish a datum from a specified datum feature.

Datum Feature Simulator (Physical)
The physical boundary used to establish a simulated datum from a specified datum feature.

Datum Target
A specified point, line, or area on a part used to establish a datum.

Datum Target (Movable)
In geometric dimensioning and tolerancing it is a symbol that indicates that a datum target is not fixed at it basic location and is free to translate.

Datum Translation
In geometric dimensioning and tolerancing it is a symbol that indicates that a datum feature simulator is not fixed at its basic location and shall be free to translate.

Dead-Man Control
A control that requires a continuous pressure or contact by the operator to maintain machine, equipment, component, subsystem, or system operation. Such operation automatically returns to noncritical state once operator pressure or contact is removed.

Dead Time
An interval following a response to one signal or event during which a system is unable to respond to another.

Debond
A deliberate separation of a bonded joint or interface, usually for repair or rework purposes (see also Disbond).

Debug
To examine or test a procedure, routine, or equipment for the purpose of detecting and correcting errors. To detect, locate, and remove mistakes from a routine or malfunctions from a computer.

Debugging
A process to detect and remedy inadequacies. Not to be confused with terms such as *burn-in, fault isolation*, and *screening*.

Decalcomania
(1) A process of transferring design from specially prepared paper to a surface and permanently fixing them thereto. (2) The paper on which designs are printed for transfer by this method.

Decarburization
The loss of carbon from the surface of a ferrous alloy as a result of heating in a medium that reacts with the carbon.

Decay Time
The time taken by a quantity to decay to a stated fraction of its initial value. The fraction is commonly $1/e$, where e is the base of natural logarithms ($1/e = 0.37$).

Decay Time, Pulse
The time required for the instantaneous amplitude of a pulse to go from 90 to 10% of a peak value.

Deception, Device
A device that works to confuse unfriendly signals so as to make them either unusable or misleading.

Deception, Electronic
The deliberate radiation, reradiation, reflection, or absorption of electromagnetic waves in a manner intended to mislead an enemy in the interpretation of data received by an electronic control system.

Decibel (dB)
Dimensionless unit for expressing the ratio of two values, the number of decibels being 10 times the logarithm to the base 10 of a power ratio, or 20 times the logarithm to the base 10 of a voltage or current ratio.

Decipher
To convert enciphered text to the equivalent plain text by means of a cipher system. *Note*: this does not include solution by cryptanalysis.

Declaration File
A file accompanying any set of transferred files making up a document; provides all information necessary for the successful disposition of the digital files at the destination but has no purpose beyond that function.

Decode
(1) To convert data by reversing the effect of some previous encoding. (2) To interpret a code. (3) To convert encoded text into its equivalent plain text by means of a code. *Note*: this does not include solution by cryptanalysis.

Decoder
A device for translating electrical signals into predetermined functions, as in an airborne guidance system, which accepts only properly coded guidance and control command signals.

Decoding
The process of deriving a bitmap from an octet string by translating any compression algorithm used to create the octet string.

Decoding System
A program that reads and interprets a file of the specified type, which may not have been produced locally.

Decollimation
In optics, that effect wherein a beam of parallel light rays is caused to diverge or converge from parallelism. *Note*: any of a large number of factors may cause this effect, e.g., refractive index in homogeneities, occlusions, scattering, deflection, diffraction, reflection, and refraction.

Decoy
A countermeasure device used to divert a missile or antimissile from its target.

Decrypt
To convert encrypted text into its equivalent plain text by means of a crypto-system. *Note*: the term *decrypt* covers the meanings of decipher and decode.

Dedendum
The radial distance between the pitch and root diameter cylinders or cones. This term applies to those threads having a recognized pitch cylinder or pitch cone.

Dedicated Circuit
A circuit designated for exclusive use by specified users.

Dedicated Key
A key that produces one code and is never affected by the position of either the CTL or SHIFT keys.

Default Value
(1) A predetermined, frequently used value for a data field or control entry, intended to reduce required user entry actions. (2) Attribute value that is the standard value in document interchange in the context of a given application profile.

Defect
A defect is any nonconformance of the unit or product with specified requirements.

Defective
Any nonconformance of a characteristic with specified requirements.

Defects, Critical
A defect that would result in hazardous or unsafe conditions for individuals using, maintaining, or depending on the product; or a defect that judgment and experience indicate is likely to prevent performance of the tactical function of a major end item or major part thereof.

Defects, Major
A defect, other than critical, that is likely to result in failure or to reduce materially the usability of the unit of product for its intended purpose.

Defects, Minor
A defect that is not likely to reduce materially the usability of the unit of product for its intended purpose, or a departure from established standards having little bearing on the effective use or operation of the unit.

Defense Transportation System (DTS)
The portion of the worldwide transportation infrastructure that supports Department of Defense transportation needs across the range of military operations. The DTS consists of those common-user military and commercial assets, services, and systems organic to, contracted for, or controlled by the DoD. It includes military-controlled or operated terminal facilities, Air Mobility Command controlled or arranged airlift, Military Sealift Command controlled or arranged sealift, and Government-controlled air or land transportation.

Definitive Item Levels
Definitive systems, subsystems. Centers, centrals sets, groups and units are those configurations which have a specific complement listing.

Deflection, Total
The movement of a spring from its free position to maximum operating position. In a compression spring, it is the deflection from the free length to the solid length.

Deformation
The change in shape of a specimen caused by the application of a load or force.

Degradation
A decrease in the quality of a desired signal (i.e., decrease in the signal-to-noise ratio or an increase in distortion), or an undesired change in the operational performance of equipment as the result of interference.

Delamination
(1) A separation between plies within the base material, or between the base material and the conductive foil, or both. (2) The separation of the layers of material in a laminate. This may be local or may cover a large area of the laminate. It may occur at any time in the cure or subsequent life of the laminate and may arise from a wide variety of causes.

Delay Fault
A fault in a digital device such that switching occurs to the proper level but does so outside a specified time interval. Also referred to as a *fault* or *delay*.

Delay Line
A transmission line or equivalent device designed to delay a signal for a predetermined length of time.

Delay Logic Function
A sequential logic function wherein each state transition of the input signal causes a single delayed state transition at the output.

Delivery
Data items are deemed to be delivered either when they are electronically transmitted to the Government or when they are made available for Government access, and the contractor has given notice of delivery to the Government.

Demarcation Point (DEMARC)
That point at which operational control or ownership changes from one organizational entity to another.

Demodulation
The process wherein a signal resulting from previous modulation is processed to derive a signal having substantially the characteristic of the original modulating signal.

Demonstrate
When used relative to test and evaluation, *demonstrate* implies a qualitative test that does not require comparison of test results to an applicable requirement.

Demonstrated
That which has been measured, within specified confidence limits, by the use of objective evidence gathered under specified conditions.

Demonstration and Validation Phase
The period when selected candidate solutions are refined through extensive study and analyses, hardware development (if appropriate), and test and evaluations.

Demultiplex
To separate two or more signals previously combined by compatible multiplexing equipment.

Denier
A term used to describe the weight of a yarn (not cotton or spun rayon), which in turn determines its physical size.

Density
In a facsimile system, a measure of the light transmission or reflection properties of an area, expressed by the logarithm of the ratio of incident to transmitted or reflected light flux. *Note*: there are many types of density that will usually have different numerical values for a given material: e.g., diffuse density, double diffuse density. The relevant type of density depends on the geometry of the optical system in which the material is used.

Departmental Control Point (DCVP)
The Departmental Control Point is the official control point within the military departmental authorized to obtain type designator nomenclature from the Department of Defense Control Point.

Department of Defense Control Point (DODCP)
The Department of Defense Control Point is the official assigning agency of type designators for the Department of Defense within this system.

Dependability
A measure of the degree to which an item is operable and capable of performing its required function at any (random) time during a specified mission profile, given item availability at the start of the mission (see also Availability).

Depolarization
A removal of factors resisting the flow of current in a cell.

Deposit
A foreign substance, which comes from the environment, adhering to a surface of a material.

Depot Maintenance
Maintenance performed on material requiring major overhaul or a complete rebuild of parts, subassemblies, and end items, including the manufacture of parts, modification, testing, and reclamation as required. Depot maintenance serves to support lower categories of maintenance by providing technical assistance and performing maintenance that is beyond their responsibility. Depot maintenance provides stocks or serviceable equipment by using more extensive facilities for repair than are available in lower-level maintenance activities.

Depot Support Equipment (DSE)
That class of equipment, excluding common hand tools, necessary to overhaul or repair and test contractor hardware to the lowest reparable unit. This includes commercial equipment as well as equipment specifically designed or built to fulfill a particular depot overhaul or repair function.

Depot Technical Data
Documentation (control manuals, supplemental data, engineering data, other technical orders, etc.) used by technicians during depot-level maintenance on aerospace vehicle equipment and support equipment. The control manual shall identify the functions (i.e., repair, install, calibrate, etc.) to be performed on the aerospace vehicle equipment and support equipment and direct the technician to the appropriate data to perform the function.

Depth of Fusion
The distance that fusion extends into the base metal from the surface melted during welding.

Derating
Derating of a part is the intentional reduction of its applied stress, with respect to its rated stress, for the purpose of providing a margin between the applied stress and the demonstrated limit of the part's capabilities. Maintaining this derating margin reduces the occurrence of stress-related failures and helps ensure the part's reliability.

Desensitization
A reduction in receiver sensitivity to the desired signal due to the presence of a high-level undesired signal that overloads the receiver circuits.

Desiccant
A dehydrating agent. A material that will absorb moisture by physical or chemical means.

Desiccant, Activated
A desiccant that has been physically treated, by means such as heating, etc., to produce the maximum capacity for absorption.

Design Activity
A design activity is an activity having responsibility for the design of an item. The activity may be a Government, commercial, or nonprofit organization.

Design Activity, Current
The activity currently having responsibility for design, drawing preparation, and maintenance. Current design activity could be the Original Design Activity or a new activity that accepted transfer of responsibility from another Government activity or contractor.

Design Activity Identification (DAI)
A unique identifier that distinguishes one design activity or organization from another design activity or organization. Examples of activity identification include activity name, activity name and address, and CAGE Code.

Design Activity, Original
The activity that had original responsibility for the design of an item and whose drawing and CAGE code appear in the title block of the drawing.

Design Adequacy
The probability that a system or equipment will successfully accomplish its mission, given that the system is operating within design specifications.

Design Analysis
Engineering analysis of the means by which an item is able to respond to the demands of its missions; the basis for classification of characteristics.

Design Disclosure Drawings
A drawing or set of drawings and associated data that delineates the detailed engineering requirements of an end product necessary for the fabrication, assembly, inspection, and test of the item.

Design Fault
A fault due to inadequate hardware or software design. Also referred to as *Fault, design.*

Design for Testability
A design process or characteristic thereof such that deliberate effort is expended to assure that a product may be thoroughly tested with minimum effort, and that high confidence may be ascribed to test results.

Design Information Drawing
A drawing that serves as the basis for developing complete details of a design or provides a graphic summary of the functional characteristics of the design.

Design Life
The operational life of equipment (to include storage life, installed life in a nonoperating mode, and operational service life), after which the equipment will be replaced or recertified. It is the responsibility of the program/project to determine recertification requirements, which may include refurbishment, analysis, or test.

Design Margin
The additional performance capability above required standard basic system parameters that may be specified by a system designer to compensate for uncertainties.

Design Maturity
The extent to which the final design or configuration of an item has been defined by the engineering process. For example, the design of a sheet metal cover having all holes in its mounting hole pattern fully dimensioned and tolerance for final size, location, and orientation would be considered to be more mature than the design of a similar cover having its mounting hole pattern defined as "drill at assembly."

Design Objective
Any desired performance characteristic for communication circuits and equipment that is based on engineering judgment but, for a number of reasons, is not considered feasible to establish as a system standard at the time the standard is written. *Note*: examples of reasons for designating a performance characteristic as a Design Objective rather than as a standard are as follows: (a) It may be bordering on advancement in the state of the art. (b) The requirement may not have been fully confirmed by measurement or experience with operating circuits. (c) It may not have been demonstrated that it can be met considering other constraints such as cost and size. A design objective must be considered as guidance for DoD agencies in preparation of specification for development or procurement of new equipment or systems and must be used if technically and economically practicable at the time such specifications are written.

Design Review
A formal, documented, comprehensive, and systematic examination of a design to evaluate the design requirements and the capability of the design to meet these requirements and to identify problems and propose solutions.

Desktop Publishing (DTP)
The typeset quality that can now be produced using a computer and laser printer for such items as brochures, newsletters, reports, etc.

Destination System
The computer hardware and software system receiving transferred data.

Destructive Physical Analysis (DPA)
A destructive physical analysis (DPA) is a systematic, logical, detailed examination of parts during various stages of disassembly, conducted on a sample of completed parts from a given lot, wherein parts are examined for a wide variety of design, workmanship, and processing problems that may not show up during normal screening tests. The purpose of these analyses is to maintain configuration control and determine those lots of parts delivered by a vendor that have anomalies or defects such that they could, at some later date, cause degradation or catastrophic failure of a system.

Destructive Testing
(1) Prolonged endurance testing of equipment or a specimen until it fails in order to determine service life or design weakness. (2) Testing in which the preparation of the test specimen or the test itself may adversely affect the life expectancy of the unit under test (UUT) or render the sample unfit for its intended use.

Detail Drawing
A detail drawing provides the complete end-product definition of the part or parts depicted on the drawing. A detail drawing establishes item identification of each part depicted thereon.

Detailed Design Data
Technical data that describes the physical configuration and performance characteristics of an item or component in sufficient detail to ensure that an item or component produced in accordance with the technical data will be essentially identical to the original item or component.

Detail Specification
A specification that specifies design requirements, such as materials to be used, how a requirement is to be achieved, or how an item is to be fabricated or constructed. A specification that contains both performance and detail requirements is still considered a detail specification. Both defense specifications and program-unique specifications may be designated as a detail specification.

Detectable Failure
A failure that can be detected with 100% detection efficiency.

Detector
(1) In a radio receiver, a circuit or device that recovers the signal of interest from the modulated wave. (2) In an optical communication receiver, a device that converts the received optical signal to another form.

Deterioration
A general term describing the impairment of desired physical, chemical, mechanical, or electrical properties resulting from aging, environmental exposure, chemical or biological attack, or changes in temperature or pressure.

Deuteranope
An individual who exhibits deuteranopia, a color vision deficiency in which the green retinal photoreceptors are absent, moderately affecting red-green hue discrimination, and thus making it difficult to distinguish between colors in the red-orange-yellow-green section of the spectrum.

Developmental Configuration
(1) The contractor's design and associated technical documentation that defines the evolving configuration of a configuration item during development. It is under the developing contractor's configuration control and describes the design definition and implementation. The developmental configuration for a configuration item consists of the contractor's internally released hardware and software designs and associated technical documentation until establishment of the formal product baseline. (2) The contractor's software and associated technical documentation that defines the evolving configuration of a computer software configuration item during development. It is under the development contractor's configuration control and describes the software design and implementation. The Developmental Configuration for a computer software configuration item consists of a Software Design Document and source code listings. Any item of the Developmental Configuration may be stored on electronic media.

Development Contractor
A development contractor is a contractor responsible for the development engineering and modification of configuration items (hardware, software, or both).

Development Design Drawings
Drawings that describe the physical and functional characteristics of a specific design approach to the extent necessary to permit the analytical evaluation of the ability of the design approach to meet specified requirements and enable the development and manufacture of experimental hardware.

Development Model
A model designed to meet performance requirements of the specification or to establish technical requirements for production equipment. This model need not have the required final form or necessarily contain parts of final design. It may be used to demonstrate the reproducibility of the equipment.

Development Testing
A series of materiel tests conducted by the Army developer with or without contractor assistance to assess program technical risks, demonstrate that engineering design is complete and acceptable, determine the extent of the design risks, determine specification compliance, and assess production requirements.

Deviation
(1) A specific written authorization, granted prior to the manufacture of an item, to depart from a particular requirement of an item's current approved configuration documentation for a specific number of units or a specified period of time. (A *deviation* differs from an engineering change in that an approved engineering change requires corresponding revision of the item's current approved configuration documentation, whereas a deviation does not). (2) A variation from an established dimension, position, standard, or value. In ISO usage, the algebraic difference between a size (actual, maximum, or minimum) and the corresponding basic size. The term *deviation* does not necessarily indicate an error.

Deviation from Normal, Allowable
Changes in indication which are acceptable during a susceptibility test, provided they do not deviate beyond the tolerance given in the individual equipment specification.

Dewetting
A condition that results when molten solder has coated a surface and then receded, leaving irregularly shaped mounds of solder separated by areas covered with a thin solder film; base metal is not exposed.

Dezincification
- see PARTING.

Diagnostic Program
A computer program that recognizes, locates, and/or explains (a) a fault in equipment, networks, or systems, (b) a predefined error in input data, or (c) a syntax error in another computer program.

Dialogue
A structured series of interchanges between a user and a computer terminal. Dialogues can be computer initiated, e.g., question and answer, or user initiated (e.g., command languages).

Diametral Pitch
The quotient of the total number of teeth in the circumference of the work divided by the basic blank diameter. In the case of the tool, it would be the total number of teeth in the circumference divided by the nominal diameter. In knurl, the diametral pitch and number of teeth are always measured in a transverse plane perpendicular to the axis of rotation for diagonal as well as straight knurls and knurling.

Diazo
Light-sensitive component of diazo-type materials that reacts with couplers to form diazo dyes.

Diazo Chrome
A diazo-type reproduction made on a transparent plastic sheet and used principally for projection and color proofing.

Diazo Paper
A reproduction paper that depends on the light sensitivity of organic dyes of the diazo type. Development is accomplished by ammonia fumes or by applying an appropriate developing solution to the face of the sheet.

Diazo Print
A diazo-type reproduction made on any one of several bases, such as paper, cloth, or plastic sheet.

Diazo Processor
A machine designed to expose and develop diazo-type materials.

Dichotic
The condition in which the sound stimulus presented at one ear differs from the sound stimulus presented at the other ear. The stimulus may differ in sound pressure, frequency, phase, time, duration, or bandwidth.

Dichroic
Exhibiting the quality of dichroism. Having two colors (hues). (1) Pertaining to certain crystals that show different colors when viewed in different direction by transmitted light. (2) Pertaining to a change of hue with the thickness or concentration of a colored medium, especially certain glasses and dye solutions that are blue or green in low densities and red when the density is high.

Dichroism
As applied to anisotropic materials, such as certain crystals, this term refers the selective absorption of light rays vibrating in one particular plane relative to the crystalline axes, but not those vibrating in a plane at right angles thereto. As applied to isotropic materials, this term refers to the selective reflection and transmission of light as a function of wavelength

regardless of its plane of vibration. The color of such materials, as seen by transmitted light, varies with the thickness of material examined. An alternative term for this phenomenon might be *polychromatism*.

Dickey Fix
Technique designed to protect a receiver from fast sweep jamming.

DID
- see DATA ITEM DESCRIPTION.

Die
Any of various tools or devices for imparting a desired shape, form, or finish to a material or for impressing an object or material.

Die Closure
Allowable part thickness variation caused by inconsistent mating of opposing segments of a mold or die.

Dielectric
(1) Any substance in which an electric field may be maintained with zero or near-zero power dissipation. (2) Any insulating medium that intervenes between two conductors and permits electrostatic attraction and repulsion to take place across it. (3) A material having the property that energy required to establish an electrical field is recoverable in whole or in part, as electric energy.

Dielectric Absorption
That property of an imperfect dielectric whereby there is an accumulation of electric charges within the body of the material when it is placed in an electrical field.

Dielectric Breakdown
A complete failure of a dielectric material characterized by a disruptive electrical discharge through the material due to a sudden and large increase in voltage.

Dielectric Constant
(1) The ratio of the capacity of a condenser having a dielectric constant between the plates to that of the same condenser when the dielectric is replaced by a vacuum; a measure of the electrical charge stored per unit volume at unit potential. (2) That property of a dielectric which determines the electrostatic energy stored per unit volume for unit potential gradient.

Dielectric Lens
A lens made of dielectric material that refracts radio waves in the same manner that an optical lens refracts light waves.

Dielectric Strength
The maximum voltage that a dielectric can withstand under specified conditions without resulting in a voltage breakdown (usually expressed as volts/unit dimension).

Differential Aeration Cell
A special type of concentration cell in which potential differences are established due to local differences in dissolved oxygen content.

Diffusion Welding (DFW)
A solid state welding process wherein coalescence of the faying surfaces is produced by the application of pressure and elevated temperatures. The process does not involve macroscopic deformation or relative motion of the parts. A solid filler metal may or may not be inserted.

Dig
A short scratch whose width is sufficient to be measured.

Digit
A symbol, numeral, or graphic character that represents an integer, e.g., one of the decimal characters "0" through "9," or one of the binary characters "0" or "1." *Note*: in a given numeration system, the number of allowable different digits, including zero, is always equal to the radix (base).

Digital Data
(1) Data stored on a computer system that employs a display on which the user and the computer interact to create entities for the production of layouts, drawings, numerical control tapes, or other engineering data. (2) Data represented in discrete discontinuous form as contrasted with analog data represented in continuous form.

Digital Driver
The output stage of a digital data generator.

Digital Error
A single-digit inconsistency between the signal actually received and the signal that should have been received.

Digital Switching
A process in which digital signals are switched without converting them to or from analog signals.

Digital-to-Analog Converter
(1) A device that converts a digital input signal to an analog output signal carrying equivalent information. (2) A functional unit that converts data from a digital representation to an analog representation.

Digital Word
A binary number whose bit length corresponds to that typical of memory or basic arithmetic operations.

Digitize
To convert an analog signal to a digital signal carrying equivalent information.

Digitizer
A device that provides input coordinate data by scanning or pointing. Digitizers are used in converting a display, drawing, or image to digital form.

Dimension
A numerical value expressed in appropriate units of measure and indicated on a drawing and in other documents along with lines, symbols, and notes to define the size or geometric characteristic, or both, of a part or part feature.

Dimension, Basic
(1) A numerical value used to describe the theoretically exact size, profile, orientation, or location of a feature or datum target. (2) A theoretically exact dimension.

Dimension, Reference
A dimension, usually without tolerance, used for information purposes only. A reference dimension is a repeat of a dimension or is derived from other values shown on the drawing or

on related drawings. It is considered auxiliary information and does not govern production or inspection operations.

Dimensional Stability
A measure of dimensional change caused by such factors as temperature, humidity, chemical treatment, age, or stress.

Dimension Lines
A dimension line, with its arrowheads, shows the direction and extent of a dimension. Numerals indicate the number of units of a measurement.

Dimetric Projection
A diametric projection is an axonometric projection in which two axes of the object make equal right angles with the plane of projection and the third axis makes a different angle with the plane of projection. Two of the angles between axes are equal; the third angle is unequal.

Dimpling
Stretching a relatively small, shallow indentation into sheet metal. Stretching metal into a conical flange for use of a countersink rivet or screw.

Diode, Current-Regulator
A diode that limits current to an essentially constant value over a specified voltage range.

Diode, Monolithic and Multiply Array
Consists of several diodes fabricated in a single monolithic chip. Monolithic arrays allows diode interconnection to form a desired circuit, whereas multiple arrays are restricted to a given circuit configuration such as a common anode circuit or a common cathode circuit.

Diode, Photo
A diode that is responsive to radiant energy.

Diode, Semiconductor
A semiconductor device having two terminals and exhibiting a nonlinear voltage-current characteristic.

Diode, Transient Voltage Suppressor
Transient voltage suppressors are characterized by two zener diodes oriented back to back and are capable of high-voltage transient suppression.

Diode, Tuning
A varactor diode used for RF tuning, including functions such as automatic frequency control and automatic fine tuning.

Diode, Varactor
A two-terminal semiconductor device in which use is made of the property that its capacitance varies with the applied voltage.

Diode, Voltage-Reference
A diode that is normally biased to operate in the breakdown region of its voltage-current characteristic, and that develops across its terminals a reference voltage of specified accuracy when biased to operate throughout a specified current and temperature range.

Diode, Voltage-Regulator
A diode that is normally biased to operate in the breakdown region of its voltage-current characteristic and that develops across its terminals an essentially constant voltage throughout a specified current range.

Dip
A hollow in an optical surface.

Dip Brazing (DB)
A brazing process in which the heat required is furnished by a molten chemical or metal bath. When a molten chemical bath is used, the bath may act as a flux. When a molten metal bath is used, the bath provides the filler metal.

Dip Soldering (DS)
A soldering process in which the heat required is furnished by a molten bath, which provides the solder.

Direct Impingement
Passing cooling air over parts without the use of cold plates or heat exchangers.

Directional Antenna
An antenna in which the radiation pattern is not omnidirectional, i.e., a nonisotropic antenna.

Directional Coupler
A transmission coupling device for separately sampling (through a known coupling loss) either the forward (incident) or the backward (reflected) wave in a transmission line.

Direct Positive
A positive image obtained directly from another positive image without the use of a negative.

Direct Positive Process
Any photographic process that yields a positive image of the original scene or photographic sound track without the formation of a negative image as an intermediate step. Distinguished from *Negative-Positive Process, Reversal Process.*

Disbond
An area within a bonded interface between two adherents in which an adhesion failure or separation has occurred. It may occur at any time during the life of the structure and may arise from a wide variety of causes. Also, colloquially, an area of separation between two lamina in the finished laminate (in this case the term *delamination* is normally preferred) (see also Debond).

Discrete Code
A bar code in which the inter-character gap is not part of the code and is allowed to vary dimensionally within wide tolerance limits.

Discrete Component
A separate part of a printed board assembly that performs a circuit function, e.g., a resistor, a capacitor, a transistor, etc.

Discrete Wiring Board
A base material upon which discrete wiring techniques are used to obtain electrical interconnections.

Discrete Wiring Board Assembly
An assembly that uses a discrete wiring board for component mounting and interconnecting purposes.

Disk Operating System (DOS)
This is the program that tells the computer how to operate.

Display Elements
In computer graphics, the basic building symbols for an application used to construct display images, e.g., points, line segments, and characters.

Display Format
The organization of different types of data in a display, including information about the data such as labels and other user guidance such as prompts, error messages, etc.

Display Framing
User control of display coverage by display movement, including paging, scrolling, offset, and expansion.

Display Group
A collection of display elements that can be manipulated as a unit and that may be further combined to form larger groups.

Display Image
The collection of display elements and display groups that are visually represented together on the viewing surface of a display device.

Display Menu
Option listed on a display allowing an operator to select the next action by indicating one or more choices with an input device.

Display Tailoring
Designing displays to meet the specific task needs of a user rather than providing a general display that can be used for many purposes.

Dissimilar Metals
Dissimilar metals exist when two metal specimens are in contact or otherwise electrically connected to each other in a conductive solution and generate an electric current.

Dissimilar Metal Welds
Dissimilar metal welds are required to fabricate a weldment or welded system. These welds include designed weld joints: weld buildups; overlay cladding for corrosion resistance, hard facing, and wear resistance; and weld deposited buttering.

Dissipation Factor
The ratio of the power loss to the circulating voltage.

Distribution Statement
A statement used in marking a technical document to denote the extent of its availability for distribution, release, and disclosure without need for additional approvals and authorizations from the controlling DoD office.

Divergence
The bending of rays away from each other, as by a concave or minus lens or by a convex mirror. In a binocular instrument, divergence is the horizontal angular disparity between the

images of a common object, as seen through the left and right systems. Divergence is defined as positive when the right image is to the right of the left image.

Document
A term applicable to the specifications, drawings, lists, standards, pamphlets, reports, and printed, typewritten, or other information, relating to the design, procurement, manufacture, testing, or acceptance inspection of items or services.

Document Application Profile (DAP)
The result of selecting a particular document architecture level; content architectures; a document profile level; an interchange format level; objects and attributes with classification of attributes into mandatory, non-mandatory; and defaultable, and definitions of basic, non-basic, and default attribute values, and control function parameter values.

Document Image Standard
A technical standard describing the digital exchange format of a print/display file of a report or other document.

Document Type Definition (DTD)
Rules, determined by an application, that apply Standard Generalized Markup Language (SGML) to the markup of documents of a particular type. A document type definition includes a formal specification, expressed in a document type declaration, of the element types, element relationships and attributes, and references that can be represented by markup. It thereby defines the vocabulary of the markup for which SGML defines the syntax.

DoD Activity Address Code (DoDAAC)
A distinct six-position alphanumeric code assigned to identify specific units, activities, or organizations.

Don't Care State
A portion of primary input or output patterns created for a UUT that are not assigned specific values.

Doppler Effect
The phenomenon evidenced by the change in the observed frequency of a sound or radio wave caused by a time rate of change in the effective length of the path of travel between the source and the point of observation.

Dot Matrix Printer
A type of printer that creates text characters with a series of closely spaced dots.

Double Sided Board
A printed board with a conductive pattern on both sides.

Down Conductor, Lightning
The conductor connecting the air terminal or overhead ground wire to the earth electrode subsystem.

Down the Slat
Vernacular expression for a successful flight of a missile down the test range and within the left and right range limits (parallel lines previously plotted on the range plotting board) established by range safety personnel.

Downtime
That portion of calendar time when the item cannot perform its intended function.

Downwash
Aerodynamic interference from the wings of an aircraft or missile that can strike the aft tail surfaces in such as way as to cause positive trim angles of attack, even in cases where the wings are located at the vehicle's center of gravity.

Draft
The taper applied to selected surfaces to aid in the removal of a part from a die or pattern from a mold. Draft adds mass to the part unless otherwise specified.

Drag Soldering
A process whereby supported, moving printed circuit assemblies or printed wiring assemblies are brought in contact with the surface of a static pool of molten solder.

Drawing (Engineering)
An engineering document or digital data file that discloses (directly or by reference), by means of graphic or textual presentations, or combinations of both, the physical and functional requirements of an item.

Drawing Form
A sheet of drafting material displaying the basic format features such as title block, general tolerance blocks, and margins.

Drawing Format
The arrangement and organization of information within a drawing. This includes such features as the size and arrangement of blocks, notes, list, revision information, and the use of optional or supplemental blocks.

Drift
In guidance, the gradual lateral deviation of a missile away from the desired trajectory, due to misalignments, electrical biases, or crosswinds.

Drift Rate
The amount of lateral deviation of a missile away from a desired trajectory, per second or any other unit of time.

Drill Tape
A tape file that contains data for running numerically controlled (NC) drilling machines.

Driver
Circuit providing energy to the light source and the modulator circuit. It is designed to modulate the light source with the desired signal, maintain light efficiency by controlling bias current, and protect the light source by limiting the bias current and, in some cases, controlling the temperature of the light source.

Driver Program
Superfluous (throw-away) code needed to perform the unit testing and lower levels of integration testing in a bottom-up software development effort.

Drivers
Computer-based informal testing techniques that are used in conjunction with other techniques. Drivers are used almost exclusively during the system implementation phase of a software development project.

Drogue Parachute
(1) A type of parachute attached to a missile to slow it down (also called *deceleration parachute* or *drag parachute*). (2) A small parachute specifically used to pull a larger parachute out of stowage.

Drogue Recovery
A recovery system for stabilizing and decelerating payloads, warheads, or booster section from missiles in the atmosphere so that larger recovery parachutes can be deployed at lower altitudes within specified opening shock constraints.

Dross
Oxide and other contaminants that form on the surface of molten solder.

Dual Dimensions
A former practice that included linear dimensions in views of a drawing in both metric and inch-pound units.

Dual Indication
The inclusion, in test or on instrumentation and gaging, of a quantity (characteristic or dimension) in both metric and inch-pound units.

Dual In-line Package (DIP)
A component that terminates in two straight rows of pins or lead wires.

Dual Port
An architectural implementation that allows ATE hardware resource sharing between two ATE interfaces, which may be used for testing different UUTs.

Duct Sheet
Coiled or flat sheet in special tempers, widths and thicknesses, suitable for duct applications.

Dummy Connector, Receptacle
An item specifically designed to mate with a plug connector to perform one or more special functions. It does not have provisions for attaching a cable. Dummy receptacles normally have no inserts or contacts present and are usually simple receptacle shells.

Dummy Load
A device or any electronic circuit that provides a simulation of the normal input or output of a circuit or a system under test.

Dunnage
Lumber, strapping, nails, or other material used to secure and protect lading.

Duplex Circuit
A circuit that permits simultaneous transmission in both directions.

Duplexer
A device that permits the simultaneous use of a transmitter and a receiver in connection with a common element such as an antenna system.

Duplicate Original
A replica of an engineering drawing or digital data file created to serve as the official record of the item when the original has been lost.

Duty Cycle
(1) The ratio of the sum of all pulse durations to the total period, during a specified period of continuous operation. (2) Time intervals of starting, running, stopping, and idling intermittent-duty devices. (3) Ratio of working time to total time for intermittently operated devices.

Dwell Time
That period of time during which a dynamic process is halted to allow another process to occur.

Dynamic Model
A model of a missile or other vehicle reproduced in a scale (dimensions, weight, and moments of inertia) proportionate to the original.

Dynamic Range
(1) In a transmission system, the ratio of the overload level to the noise level of the system, usually expressed in decibels. (2) The ratio of the specified maximum level of a parameter (e.g., power, voltage, frequency, or floating point number representation) to its minimum detectable or positive value, usually expressed in decibels.

Dynamic Test
A test of one or more of the signal properties or characteristics of an equipment or any of its constituent items performed such that the parameters being observed are measured and assessed with respect to a specified time aperture or response.

Dynamic Variation
A short time variation outside of steady-state conditions in the characteristics of power delivered to communication equipment.

E

Earth Electrode Subsystem
A network of electrically interconnected rods, plates, mats, or grids installed for the purpose of establishing a low-resistance contact with earth.

Earthing
The process of making a satisfactory electrical connection between the structure, including the metal skin, of an object or vehicle, and the mass of the earth to ensure a common potential with the earth.

Ease of Maintenance
The degree of facility with which equipment can be retained in, or restored to, operation. It is a function of the rapidity with which maintenance operations can be performed to avert malfunctions or correct them if they occur. Ease of maintenance is enhanced by consideration that will reduce the time and effort necessary to maintain equipment at peak operating efficiency.

Eccentricity (Screw Thread)
The distance between the axis of the pitch cylinder and either the axis of the major diameter cylinder or the axis of the minor diameter cylinder. Eccentricity is half of the concentricity.

Echo
(1) Wave that has been reflected or otherwise returned with sufficient magnitude and delay to be perceived. (2) Signal reflected by a distant target to a radar set. (3) Deflection or indication on the screen of a cathode-ray tube representing a target.

Edge Board Connector
A connector designed specifically for making removable and reliable interconnections between the edge board contacts on the edge of a printed board and external wiring.

Edge Connector
The portion of a circuit board which is used for communication of input, output and power signals between itself and the prime system.

Edge Definition
The fidelity of reproduction of a pattern edge relative to the production master.

Edge Spacing
The distance of a pattern, components, or both, from the edges of the printed board.

Effective Exhaust Velocity
The calculation of rocket engine performance limitations based on the product of the specific impulse and gravitational conversion factor; the total velocity of the exhaust stream after the effects of friction, heat transfer, and nonaxially directed flow.

Effective Propellant
The total propellant minus the propellant that is consumed in starting and shutdown, or that which is trapped in tanks, pumps, lines, valves, or cooling jackets.

R. Hanifan, *Concise Dictionary of Engineering: A Guide to the Language of Engineering*,
DOI 10.1007/978-3-319-07839-7_5, © Springer International Publishing Switzerland 2014

Effective Temperature
An arbitrary index that combines into a single value the effect of temperature, humidity, and air movement on the sensation of warmth or cold felt by the human body. The numerical value is that of the temperature of still, saturated air that would induce an identical sensation.

Effective Thread
The effective (or *useful*) thread includes the complete thread and those portions of the incomplete thread that are fully formed at the root but not at the crest (in taper pipe threads, this includes the so-called black crest threads), thus excluding the vanish thread.

Effective Thrust
The theoretical thrust, in a rocket motor or engine, minus the incomplete combustion and friction flow in the nozzle.

Egg Crating
A method of dunnaging so that each unit of lading is confined in its own cell.

Elastomer
A family of plastics often used in connectors. A molded plastic component of a splice or connector that deforms slightly under pressure from the inserted fibers, resulting in alignment of the fiber ends.

Electrical Field
A vector field about a charged body. Its strength at any point is the force that would be exerted on a unit positive charge at that point.

Electrical Isolation
Separation of electrical circuits, signals, or data to preclude ambiguity, interference, or information perversion. This may be achieved through physical isolation or by any property that distinguishes one electrical signal from all others (for example, time, phase, amplitude, or frequency).

Electrically Conductive Material
Electrically conductive material refers to material whose electrical conductivity is greater than or equal to 0.5% of the electrical conductivity of copper.

Electrically Continuous Path
An electrically continuous path is a path of electrically conductive material not containing a resistive element.

Electrically Continuous Surface
An electrically continuous surface is a surface of electrically conductive material not containing a resistive element.

Electrical Steel
Electrical steel is a term used commercially to designate a flat-rolled iron-silicon alloy used for its magnetic properties.

Electrical Surge Suppressor, Marine Type
A power line conditioning device to protect vulnerable electronic equipment from common mode and differential mode voltage and current transients generated in other electrically-powered onboard systems sharing electrical power lines as opposed to a separate power line/ground system in accordance with a specific installation control drawing.

Electric Propulsion
A general term encompassing all the various types of propulsion in which the propellant consists of charged electrical particles that are accelerated by electrical or magnetic fields or both; for example, electrostatic propulsion, electromagnetic propulsion, and electrothermal propulsion.

Electrical Interchangeability
The modified item's capability of operation must be equal to the basic or previous item without requiring any modifications.

Electro-chemical Solution Potential
A potential of a metal measured with respect to a standard (reference) half-cell such as a saturated calomel electrode (also known as *electrode potential*).

Electrode
An electrical conductor in contact with an electrolyte that serves as an electron acceptor or donor (see Anode and Cathode as specific examples).

Electrodeposited Foil
A metal foil that is produced by electro-deposition of the metal onto a material acting as a cathode.

Electro-deposition
The deposition of a conductive material from a plating solution with application of electrical current.

Electro-explosive Device (EED)
(1) Any electrically initiated explosive device within an electro-explosive subsystem that has an explosive or pyrotechnic output and that is actuated by the first element (initiator) or a pyrotechnic or explosive train. (2) An electrically initiated explosive device having an explosive or pyrotechnic output.

Electro-explosive Subsystem
All components of a subsystem required to actuate, control, and monitor an electrically initiated ordnance/pyrotechnic function.

Electroless Deposition
The deposition of metal from an autocatalytic plating solution without application of electrical current.

Electrolyte
A chemical substance or mixture, usually liquid, containing ions that migrate in an electric field.

Electromagnetic Compatibility (EMC)
(1) The capability of equipments or systems to be operated in their intended operational environment at designed levels of efficiency without causing or receiving degradation owing to unintentional EMI. EMC is the result of an engineering planning process applied during the life cycle of equipment. The process involves careful consideration of frequency allocation, design, procurement, production, site selection, installation, operation, and maintenance. (2) The capability of electrical and electronic systems, equipments, and devices to operate in their intended electromagnetic environment within a defined margin of safety and at design levels of performance without suffering or causing degradation as a result of electromagnetic interference.

Electromagnetic Environment
The totality of electromagnetic phenomena existing at a given location.

Electromagnetic Force
In welding, the force due to the interaction of the welding current with its own magnetic field.

Electromagnetic Interference (EMI)
(1) Any spurious external disturbance causing unwanted response in electronic equipment, or any unwanted signal emanating from the equipment; sometimes called *radio frequency interference (RFI)*. (2) Any electromagnetic disturbance that interrupts, obstructs, or otherwise degrades or limits the effective performance of electronics or electrical equipment. It can be induced intentionally, as in some forms of electronic warfare, or unintentionally as a result of spurious emissions and responses, intermodulation products, and the like.

Electromagnetic Pulse (EMP)
(1) An electromagnetic traveling wave resulting from a nuclear event. (2) A large impulsive-type electromagnetic wave generated by nuclear or chemical explosions.

Electromotive Force Series
A listing of elements according to their standard electrode potentials, with the value for hydrogen arbitrarily taken as 0.0 volts (also called *DMF series*).

Electron Flow
A movement of electrons in an external circuit connecting an anode and cathode in a corrosion cell: the current flow is arbitrarily considered to be in an opposite direction to the electron flow.

Electronic
Enemy in interpreting signals from his equipment.

Electronic Countermeasures, Passive
The conduct of such search, interception, direction finding, range estimation, and signal analysis of communication and noncommunication electromagnetic radiations as may be undertaken to permit immediate operational use of the information.

Electronic Counter-Countermeasures (ECCM)
That major subdivision of electronic warfare involving actions taken to ensure our own effective use of electromagnetic radiations despite the enemy's use of countermeasures.

Electronic Countermeasures (ECM)
That major subdivision of electronic warfare involving actions taken to prevent or reduce the effectiveness of enemy equipment and tactics employing or affected by electromagnetic radiations and to exploit the enemy's use of such radiations.

Electronic Material
Electronic materiel, from a military point of view, generally includes those electronic devices employed in data processing, detection and tracking (underwater, sea, land-based, air and space) recognition and identification, communications, aids to navigation, weapons control and evaluation, flight control, and electronic countermeasures. In every case, electronic devices are understood to include peculiar non-electronic units required to complete that individual operational function, but to exclude associated non-electronic equipment identified by other type designator systems. This includes certain applications of vehicles, hardware and non-electronic auxiliary equipment such as carrying cases.

Electronic Product Code (EPC)

An identification scheme for universally identifying physical objects via radio frequency identification tags and other means. The standardized EPC data consists of an EPC (or *EPC Identifier*) that uniquely identifies an individual object, plus an optional filter value when judged to be necessary to enable effective and efficient reading of the EPC tags. In addition to this standardized data, certain classes of EPC tags will allow user-defined data. The EPC tag Data Specifications will define the length and position of this data, without defining its content.

Electronic Serial Number (ESN)

The unique identification number embedded or inscribed on the microchip in a wireless phone by the manufacturer. The ESN is composed of two basic components, the manufacturer's code and the serial number, in accordance with TIA ESN assignment Guidelines and Procedures.

Electronic Warfare

That division of the military use of electronics involving actions taken to prevent or reduce an enemy's effective use of radiated electromagnetic energy and actions taken to ensure our own effective use of radiated electromagnetic energy. Electronic warfare includes electronic countermeasures and electronic counter-countermeasures.

Electrostatic Discharge (ESD)

A transfer of electrostatic charge between objects at different potentials, caused by direct contact or induced by an electrostatic field.

Electrostatic Discharge-Sensitive Items (ESDS)

Electronic parts having sensitive characteristics (e.g., thin-layered internal composition) and delicate, miniaturized construction that are susceptible to damage or degradation, in various degrees, from environmental field forces (electrostatic, electromagnetic, magnetic, or radioactive). This susceptibility also extends to the standard electronic modules, printed circuit boards, printed wiring boards, and circuit card assemblies containing one or more of these sensitive electronic parts.

Electrostatics

That class of phenomena which is recognized by the presence of electrical charges, either stationary or moving, and the interactions of these charges, this interaction being solely by reason of the charges themselves and their position and not by reason of their motion.

Element (Screw Thread)

Characteristic of a thread including, but not limited to, thread angles, root, crest, pitch, lead, lead angle, major, minor, and pitch diameters.

Embedded Languages

Embedding is the (semantic) extension of a programming language without altering either the existing facilities of that language or its processor, preferably writing the extension in the language itself. (An embedded extension adds semantic capability by making it easier for a user to do something that was possible already in the existing language, i.e., a subroutine.)

Embedded Software

Software resident in a system dedicated to a function other than that of digital computation in general.

Embedded Test

Test hardware and software that is physically enclosed in the end item or permanently attached to it. Any portions of the system's diagnostic capability that is an integral part of the prime system or support system. "Integral" implies that the embedded portion is physically enclosed in the prime system or permanently attached, physically or electrically.

Embedment (Potting)
A process for encasing a part or an assembly of discrete parts within a protective material that is generally over 2.5 mm thick, varies in thickness, fills the connecting areas within an assembly, and requires a mold or container to confine the material while it is hardening. Potting is an embedding process wherein the protective material bonds to the mold or container so that it becomes integral with the item.

Embrittlement
A loss of load-carrying capacity of a metal or alloy.

Emission
Electromagnetic energy propagated from a source by radiation or conduction.

Emission, Conducted
Electromagnetic emissions propagated along a power or signal conductor.

Emission Control (EMCON)
A ship operational condition in which acoustic, electromagnetic, and optical emitters, such as radars and communications equipment, are inhibited or limited.

Emission, Electromagnetic Interference
Any conducted or radiated emission that causes electromagnetic interference.

Emission, Harmonic
Electromagnetic radiation from a transmitter that is not part of the information signal, but whose frequency is an integral multiple of the carrier frequency.

Emission, Impulse
That emission produced by impulses having a repetition frequency not exceeding the impulse bandwidth of the receiver in use.

Emission, Parasitic
Electromagnetic radiation from a transmitter that is not part of the information signal or harmonically related to the carrier, caused by undesired oscillations in the circuitry.

Emission, Radiated
Desired or undesired electromagnetic energy propagated through space. Such an emission is called *radiated interference* if it is undesired.

Emission Spectrum
A power versus frequency distribution of a signal about its fundamental frequency, which includes the fundamental frequency and the associated modulation sidebands, as well as non-harmonic and harmonic emissions and their associated sidebands.

Emission, Spurious
Any electromagnetic emission on a frequency or frequencies that are outside the necessary emission bandwidth, the level of which may be reduced without affecting the corresponding transmission of information. Spurious emissions include harmonic emission, parasitic emission, and intermodulation products but exclude emissions in the immediate vicinity of the necessary emission bandwidth, which are a result of the modulation process for the transmission of information.

Encapsulation
(1) The embedment and complete envelopment of an item or assembly in a solid mass of a plastic, elastomeric, or ceramic insulating material. (2) A process for encasing a part or an assembly of discrete parts within a protective material that is generally not over 2.5 mm thick and does not require a mold or container.

Enclaving
A synergistic zoning of the combat system, hull, mechanical and electrical systems, and damage control systems into regions that, if necessary, can function independently to provide a subset of the ship's mission capabilities.

Enclosed Rack
An enclosed rack is constructed to have the capability of being completely enclosed.

Encoding
(1) A design and programming technique through which information is stored or conveyed by means of a reversible mapping from the domain in which the information exists originally into another domain. (2) The process of deriving compressed data from a bitmap by applying a compression algorithm to the bit-map.

Encoding System
A program that produces or outputs for export a file of the specified type.

End-of-Life Design Limit
The end-of-life design limits for an item are the expected variations in its electrical parameters over its period of use in its design environment. The parameter variations are expressed as a percentage change beyond the specified minimum and maximum values. Circuit design should accommodate these variations over the life of the system.

End-Product (End-Item)
An end-product is an item, such as individual part or assembly, in its final or completed state.

End Threads
  - see INCOMPLETE THREADS.

End-to-End Check
Test conducted on a completed wire and/or cable run to assure continuity.

Endurance Limit
A maximum cyclic stress level a metal can withstand without a fatigue failure.

Engineering Change
A change to the current approved configuration documentation of a configuration item at any point in the life cycle of the item.

Engineering Change Justification Code
A code that indicates the reason for a Class I engineering change.

Engineering Change Priorities
The priority (emergency, urgent, routine) assigned to a Class I engineering change that determines the relative speed at which the Engineering Change Proposal is to be reviewed, evaluated, and, if approved, ordered and implemented.

Engineering Change Proposal (ECP)
A proposed engineering change and the documentation by which the change is described, justified, and submitted to the Government for approval or disapproval.

Engineering Data
(1) Engineering documents such as drawings, associated lists, accompanying documents, manufacturer specifications and standards, or other information prepared by a design activity and relating to the design, manufacture, procurement, test, or inspection of items. (2) Technical data relating to the design, manufacture, procurement, test, or inspection of hardware items or services. Examples are drawings, associated lists, accompanying documents, manufacturer specifications, manufacturing planning documentation, and specifications prepared by a contractor or Government design activity. (3) Any technical data (whether prepared by the Government, contractor, or vendor) relating to the specification, design analysis, manufacture, acquisition, test, inspection, or maintenance of items or services. All information that contains authoritative engineering definition or guidance on material, constituent items, equipment or system practices, engineering methods, and process constitute engineering data.

Engineering Drawing
An engineering drawing is an engineering document that discloses (directly or by reference) by pictorial or textual presentations, or combinations of both, the physical and functional end product requirements of an item.

Engineering Drawing Package (EDP)
A collection of product-related engineering drawings and associated lists in accordance with Mil-Std-31000 and relating to design, manufacture, test, and inspection of an item or system. No longer an active term.

Engineering Release
An action whereby configuration documentation or an item is officially made available for its intended use.

Engineering Support Data
Engineering support data consists of text, schematics, drawings, program listings and computer generated outputs, functional flow diagrams, engineering reports (such as Test Strategy Reports), and any relevant technical information to provide for the life cycle support of an end item (end items may be either hardware or software end tests). Engineering support data is required for support at the Depot Level of maintenance and cognizant field activities.

Enter
An explicit user action that affects computer processing of user entries. For example, after typing a series of numbers, a user might press an "enter key" that will add them to a database, subject to data validation.

Enterprise Identifier (EID)
A unique identifier used to distinguish one activity or organization from another activity or organization. Examples of enterprise identifiers are: CAGE Code, Department of Defense Activity Address Code, Dun & Bradstreet's Data Universal Numbering System, NATO Cage Code, and GSI Company Prefix. An enterprise identifier code is uniquely assigned to an activity by an issuing agency registered in accordance with procedures outlined in ISO/IEC 15459-2. An enterprise may be an entity such as a design activity, manufacturer, supplier, depot, and program management office or a third party.

Entrained Water
Water condensed from the cooling air and carried along with the cooling air.

Entry
Entry is the instruction at which the execution of a routine begins. A "proper program" is one having only one entry and one exit. Additional entries imply increased complexity both in coupling and in internal functional composition. Multiple entries are prohibited in structured programming.

Envelope, Actual Mating
(1)   This term is defined according to the type of feature, as follows:

    (a)   *For an external feature.* A similar perfect feature counterpart of smallest size that can be circumscribed about the feature so that it just contacts the surface at the highest points. Example, a smallest cylinder of perfect form or two parallel planes of perfect form at minimum separation that just contact the highest points of the surface. For features controlled by orientation or positional tolerances, the actual mating envelope is oriented relative to the appropriate datum. Example: Perpendicular to primary datum plane.

    (b)   *For an internal feature.* A similar perfect feature counterpart of largest size that can be inscribed within the feature so that it just contacts the surface at the highest points. Example: A largest cylinder of perfect form or two parallel planes of perfect form at maximum separation that just contact the highest points of the surface. For features controlled by orientation or positional tolerances, the actual mating envelope is oriented relative to the appropriate datum.

Envelope, Actual Mating
(2)   The envelope is outside the material. A similar perfect feature counterpart of smallest size that can be contracted about an external feature or largest size that can be expanded within an internal feature so that it coincides with the surface at the highest points. Two types of actual mating envelopes—unrelated and related.

    (a)   *Unrelated actual mating envelope.* A similar perfect feature counterpart expanded within an internal feature or contracted about an external feature and not constrained to any datum.

    (b)   *Related actual mating envelope.* A similar perfect feature counterpart expanded within an internal feature or contracted about an external feature while constrained either in orientation or location or both to the applicable datum.

Envelope, Actual Minimum Material
This envelope is within the material. A similar perfect feature counterpart of largest size that can be expanded within an external feature or smallest size that can be contracted about an internal feature so that it coincides with the surface at the lowest points. Two types of actual minimum material envelopes—unrelated and related.

Envelope, Related Actual Minimum Material
A similar perfect feature counterpart contracted about an internal feature or expanded within an external feature while constrained in either orientation or location or both to the applicable datum.

Envelope, Unrelated Actual Minimum Material
A similar perfect feature counterpart contracted about an internal feature or expanded within an external feature and not constrained to any datum reference frame.

Envelope Drawing
An envelope drawing depicts an item in a development (privately or Government) or preproduction stage. Accordingly, features not shown on the drawing are left to the ingenuity of the producer in meeting the performance, design, and installation requirements that are indicated.

Environment
The aggregate of all external and internal conditions (such as temperature, humidity, radiation, magnetic and electric fields, shock, vibration, etc.) either natural or man-made, or self-induced, that influence the form, performance, reliability, or survival of an item.

Equipment
Any electrical, electronic, or electromechanical device or collection of items intended to operate as an individual unit and performing a singular function.

Equipment Failure
An equipment failure is the cessation of the ability to meet the minimum performance requirements of the equipment specification. Furthermore, equipment failure shall imply that the minimum specified performance cannot be restored through permissible readjustment of operational controls.

Equipment Ground Network
An electrically continuous network consisting of interconnected grounding plates and structural steel elements.

Erasable Programmable Read Only Memory (EPROM)
A solid-state memory device which, after being programmed, can be reprogrammed. Also known as Electronically Alterable Read Only Memory (EAROM).

Erase
(1) To obliterate information from any storage medium, e.g., to clear or to overwrite. (2) To remove all previous data from magnetic storage by changing it to a specified condition that may be an unmagnetized state or predetermined magnetized state.

Erosion Corrosion
A corrosion reaction accelerated by velocity and air abrasion.

Error
(1) The difference between a computed, estimated, or measured value and the true, specified, or theoretically correct value. (2) A malfunction that is not reproducible.

Error-Correcting Code
A code in which each data signal conforms to specific rules of construction so that departures from this construction in the received signals can be automatically detected, and permits the automatic correction, at the received terminal, or some or all of the errors. *Note*: Such codes require more signal elements than are necessary to convey the basic information.

Error-Detecting Code
A code in which each expression conforms to specific rules of construction, so that if certain errors occur in an expression the resulting expression will not conform to the rules of construction and thus the presence of the errors is detected. *Note*: Such codes require more signal elements than are necessary to convey the fundamental information.

Etchant
A solution used to remove, by chemical reaction, the unwanted portion of material from a printed board.

Etchback
A process for the controlled removal of nonmetallic materials from sidewalls of holes to a specified depth. It is used to remove resin smear and to expose additional internal conductor surfaces.

Etched Printed Board
A board having a conductive pattern formed by the chemical removal of unwanted portions of the conductive foil.

Etch Factor
The ratio of the depth of etch (conductor thickness) to the amount of lateral etch (undercut).

Etching
A process wherein a printed pattern is formed by chemical, or chemical and electrolytic, removal of the unwanted portion of conductive material bonded to a base.

Evaluation
The process of determining whether an item or activity meets specified criteria.

Event-Based Model
A model that describes nonsequential behavior or the interrelationships between the behaviors of several components of a software system in terms of significant occurrences during system operation (events). An event-based model can be used as a design or specification technique.

Examination
An element of inspection consisting of investigation, without the use of special laboratory appliances or procedures, of supplies and services to determine conformance to those specified requirements that can be determined by such investigations. Examination is generally nondestructive and includes, but is not limited to, visual, auditory, olfactory, tactile, gustatory, and other investigations; simple physical manipulation; gaging; and measurement.

Executable Statement
A software statement which will cause some action to be performed during test run-time.

Execution
Execution of a program causes the operations in its algorithm to be performed in some order. The order in which an algorithm's operations are performed is determined by the control structures and control statements used to organize the operations. A program is executed by a processor.

Execution Analysis
The automated monitoring of the computer-based software testing activities, collection of data from these testing activities, and subsequently predicting, by manually analyzing the data, the duration and cost of testing and the quality of the software product.

Execution Time
The actual processor time utilized in executing a program.

Exfoliation
A thick, layered growth of loose corrosion products often separating from the metal surfaces.

Exit
An exit is the place where control leaves a routing.

Expander
A device the restores the dynamic range of a compressed signal to its original dynamic range.

Expansion
The restoration of the dynamic range of a compressed signal to its original dynamic range.

Exploded Pictorial Assembly Drawings
An exploded pictorial assembly drawing shows the parts of an assembly separated but in proper position and alignment for reassembly.

Export Administration Act (EAA)
Any of the laws that have been codified at 50 U.S.C. Appendix 2401–2420. These laws were the original basis for the Export Administration Regulations (EAR) (15 CFR 368–399), which require obtaining a license from the Department of Commerce for exporting certain items and related technical data.

Export Control Laws
Any law that regulates exports from the U.S. or requires obtaining a license to make such exports.

Extension (Projection) Lines
Extension lines are used to indicate the extension of a surface or point to a location outside the part outline.

Extension of Functions
An item which extends the use of equipment beyond its assigned functions and is issued for use with that equipment only under special circumstances is considered as "used with" but "part of" that equipment.

Exterior Container
A container, bundle, or assembly that is sufficient by reason of material, design, and construction to protect unit packs and intermediate containers and their contents during shipment and storage. It can be a unit pack or a container with a combination of unit packs or intermediate containers. An exterior container may or may not be used as a shipping container.

External Conductive Interfaces (Printed Wiring Board)
An external conductive interface is considered to be the junction between the surface copper foil and the deposited or plated copper.

External Layer
An external layer is a conductor pattern or land on the surface of the composite board.

External Thread
A screw thread formed on the outside of a cylindrical or conical surface.

F

Fabrication Welds

Fabrication welds are welds required to fabricate a weldment or welded system. These welds include designed weld joints, weld buildups, overlay cladding for corrosion resistance hardfacing and wear resistance, and weld deposited buttering.

Facilities

A physical plant, such as real estate and improvements thereto, including building and equipment, that provides the means for assisting or making easier the performance of a system function.

Facility

(1) A building or other structure, either fixed or transportable in nature, with its utilities, ground networks, and electrical supporting structures. (2) A service provided by a telecommunication network or equipment for the benefit of the users or the operating administration. (3) A general term for the communication transmission pathway and associated equipment. (4) In a data protocol context, an additional item of information or a constraint encoded within the protocol data unit to provide the requested control. (5) A real property entity consisting of one or more of the following: a building, structure, utility system, pavement, and underlying land.

Facility Ground System

The electrically interconnected system of conductors and conductive elements that provides multiple current paths to the earth electrode subsystem. The facility ground system includes the equipment ground network and the equipment racks, cabinets, conduit, junction boxes, raceways, duct-work, pipes, and other normally noncurrent-carrying metal elements.

Facsimile (Fax)

A system of telecommunication for the transmission of fixed images with a view to their reception in a permanent form.

Fail Safe

The design feature of a part, unit or equipment which allows the item to fail only into a non hazardous mode.

Fail-Safe Device

A device built into a potentially hazardous piece of equipment that provides that the equipment will remain safe to friendly users even though it might fail in its intended purpose. They may be self-destructive in the event of equipment failure or may be destroyed by command if operated remotely.

Failure

The temporary or permanent termination of the ability of an entity to perform its required function.

Failure Analysis

Subsequent to a failure, the logical systematic examination of an item, its construction, application, and documentation to identify the failure mode and determine the failure mechanism and its basic cause.

Failure Coverage
The ratio of failures detected (by a test program or test procedure) to failure population, expressed as a percentage. Also referred to as percent detect.

Failure Latency
The elapsed time between fault occurrence and failure indication.

Failure Mode and Effects Analysis (FMEA)
Analytical technique that uses the potential failure modes of a process and the resulting effects to prioritize corrective actions, and for identifying the characteristics of a process that are vital to product quality.

Failure Rate
The total number of failures within an item population, divided by the total numbers of life units expended by that population, during a particular measurement interval under stated conditions.

Fall Time
The time interval of the pulse trailing edge between the instants at which the instantaneous value first reaches specified upper and lower limits of 90 and 10% of voltage amplitude.

False Alarm
A fault indicated by BIT or other monitoring circuitry where no fault exists.

Far Field
The region of the field of an antenna where the radiation field predominates and where the angular field distribution is essentially independent of the distance from the antenna.

Fatigue Resistance
Resistance to metal crystallization that leads to conductors or wire breaking from flexing.

Fault
(1) A physical condition that causes a device, a component, or an element to fail to perform in a required manner; for example, a short circuit or a broken wire, or an intermittent connection. (2) Degradation in performance due to detuning, maladjustment, misalignment, or failure of parts. (3) Immediate cause of failure (e.g., maladjustment, misalignment, defect, etc.). (4) An unintentional short circuit, or partial short circuit (usually of a power circuit), between energized conductors or between an energized conductor and ground.

Fault Coverage, Fault Detection
The ratio of failures detected (by a test program or test procedure) to failure population, expressed as a percentage.

Fault Isolation
The process of determining the location of a fault to the extent necessary to affect repair.

Fault Isolation Time
The elapsed time between the detection and isolation of a fault; a component of repair time.

Fault Resolution, Fault Isolation
The degree to which a test program or procedure can isolate a fault within an item; generally expressed as the percentage of the cases for which the isolation procedure results in a given ambiguity group size.

Fault Tolerance
The capacity of a system or program to continue operation in the presence of specified faults.

Faying Surface
The surface of a member that is in contact or in close proximity with another member to which it is to be joined.

Feathers
Feathery flaws located inside the body of glass.

Feature
(1) The general term applied to a physical portion of a part, such as a surface, hole, or slot.
(2) A physical portion of a part such as a surface, pin, hole, or slot or its representation on drawings, models, or digital data files.

Feature, Axis of
(1) A straight line that coincides with the axis of the true geometric counterpart of the specified feature. (2) The axis of the unrelated actual mating envelope of a feature.

Feature, Center Plane of
(1) A plane that coincides with the center plane of true geometric counterpart of the specified feature. (2) The center plane of the unrelated actual mating envelope of a feature.

Feature, Derived Median Plane of
An imperfect plane (abstract) that passes through the center points of all line segments bounded by the feature. These line segments are normal to the actual feature. These line segments are normal to the actual mating envelope.

Feature, Derived Median Line of
An imperfect line (abstract) that passes through the center points of all cross sections of the feature. These cross sections are normal to the axis of the actual mating envelope.

Feature of Size
One cylindrical or spherical surface, or a set of two plane parallel surfaces, each of which is associated with a size dimension.

Feature of Size Irregular
There are two types of irregular feature of size as follows:

(a) A directly tolerance feature or collection of features that may contain or be contained by an actual mating envelope that is a sphere, cylinder, or pair of parallel planes.
(b) A directly tolerance feature or collection of features that may contain or be contained by an actual mating envelope other than a sphere, cylinder, or pair of parallel planes.

Feature of Size, Regular
One cylindrical or spherical surface, a circular element, and a set of two opposed parallel elements or opposed parallel surfaces, each of which is associated with a directly tolerance dimension.

Feature, Relating Tolerance Zone Framework (FRTZF)
The tolerance zone framework that controls the basic relationship between the features in a pattern with that framework constrained in rotational degrees of freedom relative to any referenced datum features.

Feebly Magnetic Material
Feebly magnetic material is a material generally classified as nonmagnetic and whose maximum normal permeability is 2.0 or less.

Feedback
(1) The return of a portion of the output of a circuit or device to its input. (2) A timing signal used as a self-test feature in an automatic test system to verify that a control instruction has been executed before proceeding to the next control instruction.

Feedback Loop
(1) Circuitry returning a portion of its output to its input. (2) An interconnection of faults and signals such that no single test point can successfully isolate the fault location.

Ferritic
An adjective describing a body-centered cubic crystal structure found in ferrous materials.

Ferromagnetic Field
Refers to the magnetic field due to all ferromagnetic materials of and in a ship, including any stores aboard and any sweep gear aboard or streamed, while the ship is not rolling or pitching. It does not include the field of ferromagnetic materials while magnetized specifically for producing a ferromagnetic sweep field.

Fiberglass
A widely used reinforcement for plastics that consists of fibers made from borosilicate and other formulations of glass. The reinforcements are in the form of roving (continuous or chopped), yarns, mat, milled, or woven fabric.

Fiber Optics
The branch of optical technology concerned with the transmission of radiant power through fibers made of transparent materials such as glass (including fused silica) or plastic.

Fidelity
The degree to which a system, or a portion of a system, accurately reproduces, at its output, the essential characteristics of the signal impressed upon its input.

Field
(1) The volume of influence of a physical phenomenon, expressed vectorially. (2) On a data medium or in storage, a specified area used for a particular class of data, e.g., a group of character positions used to enter or display wage rates on a screen. (3) Defined logically data that are part of a record. (4) The elementary unit of a record that may contain a data item, a data aggregate, a pointer, or a link.

Field Intensity
The irradiance of an electromagnetic beam under specified conditions. *Note*: usually specified in terms of power per unit area, e.g., watts per square meter, milliwatts per square centimeter.

Field Strength
The intensity of an electric, magnetic, or electromagnetic field at a given point. *Note*: normally used to refer to the rms value of the electric field, expressed in volts per meter, or of the magnetic field, expressed in amperes per meter.

Filamentary Composites
A major form of advanced composites in which the fiber constituent consists of continuous filaments. Specifically, a filamentary composite is a laminate composed of a number of

laminae, each of which consists of a nonwoven, parallel, uniaxial, planar array of filaments (or filament yarns) embedded in the selected matrix material. Individual laminae are directionally oriented and combined into specific multiaxial laminates for application to specific envelopes of strength and stiffness requirements.

Filament Winding
A reinforced plastics process that employs a series of continuous, resin-impregnated fibers applied to a mandrel in a predetermined geometrical relationship under controlled tension.

Filiform Corrosion
A form of attack that occurs under films on metals and that appears as randomly distributed hairlines (also called *underfilm corrosion*).

Fill Bits
Unallocated bits in fixed-format header fields.

Filler
A relatively inert substance added to a material to alter its physical, mechanical, thermal, electrical, or other properties or to lower cost. Sometimes the term is used specifically to mean particulate additives.

Fillet Radius
The concave radius on the surface of a part connecting intersecting surfaces.

Fillet Weld
A weld of approximately triangular cross section joining two surfaces approximately at right angles to each other in a lap joint, T-joint, or corner joint.

Film
An optional term for sheeting having a nominal thickness not greater than 0.254 mm (0.01 inches).

Find Number or Item Number
A reference number assigned to an item in lieu of the item's identifying number on the field of the drawing and entered as a cross reference to the item number of the parts list where the item name and identification number are given. Reference designations may be used as find numbers or item numbers.

Fin Stock
Coiled sheet or foil in specific alloys, tempers, and thickness ranges suitable for manufacture of fins for heat-exchanger applications.

Firewall
Security software that can actively block unauthorized entities from gaining access to internal resources such as systems, servers, databases, and networks.

Firmware
(1) The combination of a hardware device and computer instructions and/or computer data that reside as read-only software on the hardware device. The software cannot be readily modified under program control. (2) Software that has been permanently stored in hardware (specifically in non-volatile memory). It has qualities of both software and hardware.

Firmware Drawing
The drawing describing the programming requirements of a single programmed device.

First Angle Projection
First angle projection places the object between the observer and the plane of projection. This method of projection used in some countries.

First Article
A part or assembly manufactured prior to the start of production for the purpose of assuring that the manufacturer is capable of manufacturing a product that will meet the requirements.

First Article Testing
Testing and evaluating the first article for conformance with specified contract requirements before or in the initial stage of production.

Fission
Fission is the process whereby the nucleus of a particular heavy element such as uranium 235 or plutonium 239 splits into (generally) two nuclei of lighter elements, with the release of substantial amounts of energy.

Fit
(1) The ability of an item to physically interface or interconnect with or become an integral part of another item. (2) *Fit* is the general term used to signify the range of tightness or looseness that may result from the application of a specific combination of allowances and tolerances in the design of mating parts.

Fixed Effect
A common systematic (nonrandom) shift in a group of measurements due to a controlled level change of a treatment or condition.

Flameout
The extinguishment of the flame in a reaction engine. See Burnout.

Flame Retardant Resin
A resin compounded with certain chemicals to reduce or eliminate its tendency to burn. For polyethylene and similar resins, chemicals such as antimony trioxide and chlorinated paraffin's are useful.

Flammability
Flammability is a complex characteristic which combines ease of ignition, surface flammability, heat contribution, smoke production, fire gasses, and fire endurance. Flammability is a function of chemical composition, physical configuration, temperature, availability of oxygen, and retardants or additives.

Flash
(1) Excess material that results from leakage between mating surfaces of a mold or die.
(2) The molten metal that is expelled or squeezed out by the application of pressure and solidifies around the weld.

Flash Extension
Allowable flash remnant.

Flash Coat
A thin coating usually less than 0.05 mm (0.002 inches) in thickness.

Flashover
A disruptive discharge around or over the surface of a solid insulator.

Flatness
The condition of a surface having all elements in one plane. A flatness tolerance specifies a tolerance zone defined by two parallel planes within which the surface must lie.

Flaws
Flaws are unintentional irregularities that occur at one place or at relatively infrequent or widely varying intervals on the surface. Flaws include such defects as cracks, blowholes, inclusions, checks, ridges, scratches, etc. Unless otherwise specified, the effect of flaws shall not be included in the roughness average measurements. Where flaws are to be restricted or controlled, a special note as to the method of inspection should be included on the drawing or in the specifications.

Flexible Printed Wiring
A random arrangement of printed wiring utilizing flexible base material with or without flexible cover layers.

Flexural Failure
A conductor failure caused by repeated flexing. It is indicated by an increase of resistance measured for a specified time.

Flickered
A graphical representation in which symbols are used to represent such things as operations, data, flow direction, and equipment for the definition, analysis, or solution of a problem.

Flight Envelope
A pilot of velocity versus altitude that depicts maximum and minimum velocity capabilities.

Flight Mach Number
A free-stream mach number measured in flight, as distinguishable from the number measured in a wind tunnel environment.

Floppy Disk
A flexible magnetic disk enclosed in a container.

Flow Welding
Welding process that produces coalescence of metals by heating them with molten filler metal poured over the surface to be welded until the welding temperature is attained and until the required filler metal has been added. The filler metal is not distributed in the joint by capillary attraction.

Flush Field
A field added to the header and data packets designed to reset the forward error correction (FEC) encoder.

Flutter
The vibrating and oscillating movement of a control surface caused by aerodynamic forces acting upon the surface with elastic or inertial characteristics.

Flux
(1) A chemically active compound that is capable of promoting the wetting of metals with solder. (2) The rate of flow of energy across or through a surface.

F-Number
The ratio of the equivalent focal length of an objective to the diameter of its entrance pupil.

Fogging
A reduction in the luster of a metal surface as the formation of a thin film of corrosion products.

Foil
A rolled product rectangular in cross section of thickness less than 0.006 inches. In Europe, foil is equal to and less than 0.20 mm (0.0079 inches).

Font Size to Character Height
Fonts are sized in "points", which describes measure from the top of the capital letters to the bottom of lower case letters with descenders (e.g., "j" or "y"). One point is 0.35 mm (0.0139 inches).

Foot-Candle
A unit of illuminance equal to one lumen incident per square foot. The illuminance (formerly called *illumination*) of a surface placed one foot from a light source having a luminous intensity of one candle.

Forced Air Cooling
The dissipation of heat to cooling air, including ram air, supplied by a source with sufficient pressure to flow through the unit.

Forge Welding
A solid state welding process wherein coalescence is produced by heating and by applying pressure or blows sufficient to cause permanent deformation at the interface.

Forging
A part created by plastically deforming metal. Also, the process by which metal is plastically deformed to a desired shape.

Forging Plane
A plane perpendicular to the forging direction. It normally coincides with the principal mating faces of a set of dies.

Forging Stock
A wrought or cast rod, bar, or other section suitable for forging.

Form
The shape, size, dimensions, mass, weight, and other visual parameters that uniquely characterize an item. For software, *form* denotes the language and media.

Formal Qualification Review (FQR)
The test, inspection, or analytical process by which products at the end item or critical item level are verified to have met specific procuring activity contractual performance requirements (specification or equivalent). This review does not apply to requirements verified at FCA.

Formal Qualification Testing
A process that allows the contracting agency to determine whether a configuration item complies with the allocated requirements for that item.

Format
Arrangement of bits or characters within a group, such as a word, message, or language.

Form of Thread
The form of a thread is its profile in an axial plane for a length of one pitch of the complete thread.

Form Thread
The diameter at the point nearest the root from which the flank is required to be straight.

Fortran (Formal Translation)
A programming language primarily used to express computer programs by arithmetic formulas.

Fracture-Critical
Classification of hardware where a crack could lead to a failure that results in serious personnel injury, damage to flight hardware, loss of mission, or major damage to a significant ground asset.

Fracture Toughness
A mechanical property characteristic of a metal that determines its resistance to brittle fracture.

Free Aperture
 - see APERTURE, CLEAR.

Free Caustic
An amount of alkalinity of a solution measured in terms of pH values in excess of 7 (pH 7 is neutral, e.g., water).

Free State Variation
A term used to describe distortion of a part after removal of forces applied during manufacture. This distortion is principally due to the weight and flexibility of the part and the release of internal stresses resulting from fabrication.

Free State (F)
A symbol used in geometric dimensioning and tolerancing used for feature references subject to free-state variation. The symbolic means of indicating that the geometric tolerance or datum feature applies in its "free state". When the symbol is applied to a tolerance in the feature control frame, it shall follow the stated tolerance and any modifier. When the symbol is applied to a datum feature reference, it shall follow that datum feature reference and any modifier.

Free Text
Human-readable information other than what is encoded in the machine-readable medium. Free text includes applied data and information not associated with machine-readable information if present.

Frequency Modulation (FM)
The form of angle modulation in which the instantaneous frequency of a sine wave carrier is caused to depart from the carrier frequency by an amount proportional to the instantaneous value of the modulating signal.

Frequency, Reference
A frequency having a fixed and specified position with respect to the assigned frequency. The displacement of this frequency with respect to the assigned frequency has the same absolute value and sign that the displacement of the characteristic frequency has with respect to the center of the frequency band occupied by the emission.

Frequency, Tolerance
The maximum permissible departure by the center frequency of the frequency band occupied by an emission from the assigned frequency, or by the characteristic frequency of an emission, from the reference frequency.

Fretting Corrosion
A form of deterioration caused by repetitive friction between two sliding surfaces and accelerated by a conjoint corrosive action.

Friction Welding (FRW)
A solid state welding process wherein coalescence is produced by the heat obtained from mechanically induced sliding motion between rubbing surfaces. The work parts are held together under pressure.

Fringe
An interference band, such as Newton's ring.

From-To List
Written wiring instructions in the form of a list indicating termination points.

Full Duplex
The property of a channel that allows simultaneous communications in either direction.

Full Form Thread
 - see COMPLETE THREAD.

Full Indicator Movement (FIM)
The total movement of an indicator when appropriately applied to a surface to measure its variations.

Full Mission Capable
A status code meaning that the system or equipment has all systems working that are needed to perform all of its primary missions.

Full Scale Development Phase
(1) The period when the system and the principal items necessary for its support are designed, fabricated, tested, and evaluated. (2) The third phase in the materiel acquisition process. During this phase, a system, including all items necessary for its support, is fully developed and engineered, fabricated, tested, and initially type classified. Concurrently, nonmaterial aspects required to field an integrated system are refined and finalized.

Fully Hardened
(1) The most stringent performance requirements for service and combat conditions. (2) The most stringent performance criterion for a requirement. Fully hardened equipment may be COTS, ruggedized, or militarized equipment. Fully hardened is not synonymous with *fully militarized.*

Function
Refers to features that affect the ability of an item to perform its intended purpose. Functional performance may be discrete or continuous in time, active or passive, or measurably by attributes or variables.

Functional Baseline (FBL)
(1) The approved documentation describing a system's or item's functional verification required to demonstrate the achievement of those specified characteristics. (2) The initially approved documentation describing a system's functional characteristics and verification required to demonstrate the achievement of those specified functional characteristics.

Functional Characteristics
Quantitative performance parameters and design constraints, including operational and logistic parameters and their respective tolerances. Functional characteristics include all performance parameters, such as range, speed, lethality, reliability, maintainability, and safety.

Functional Configuration Audit (FCA)
The formal examination of functional characteristics of a configuration item, prior to acceptance, to verify that the item has achieved the requirements specified in its functional and allocated configuration documentation.

Functional Configuration Documentation (FCD)
The documentation describing the system's functional, performance, interoperability, and interface requirements and the verifications required to demonstrate the achievement of those specified requirements.

Functional Designation
Words, abbreviations, or meaningful number or letter combinations, usually derived from the function of an item (for example, slew, yaw), used on drawings, instructional material, and equipment to identify an item in terms of its function. (A functional designation is neither a reference designation nor a substitute for it.)

Functional Gaging
The practice of using threaded gages of perfect form and maximum or minimum material limits to determine functional size conformance.

Functional Interchangeability
(1) A condition in which a part or unit, regardless of its physical specifications, can perform the specific functions of another part or unit. (2) The modified item must be capable of performing, without additional assistance, all the operational capabilities of the basic or previous item.

Functional Marking
The symbols, letters, numbers, and similar marking applied to indicate polarity, circuitry, and similar functional characteristics.

Functional Requirements
Parameters related to the ability of the equipment to perform its intended mission. Examples are frequency, bandwidth, and so forth.

Functional Standard
A document that establishes and defines requirements for management, design processes, procedures, practices, methods, and data applicable to the creation of data products.

Functional Test
(1) A test that determines whether the unit under test is functional properly. The operational environment (such as stimuli and loads) can be either actual or simulated. (2) A test intended to exercise an identifiable function of a system. The function is tested independent of the hardware implementing the function.

Functionally Required Hardware
Hardware included in system design to satisfy any requirement other than nuclear hardening.

Fungi
Any of a large group of thallophytic organism described as molds, mildews, mushrooms, yeasts, or any organism absorbing nutrition from living or dead organic materials.

Fungicidal
Capable of killing fungi.

Funginert
Neither destroying nor supporting the growth of fungi.

Fungi Nutrient
Providing sustenance for fungi.

Fungus Resistant
Unaffected by fungi.

Furnace Brazing (FB)
A brazing process in which the heat required is obtained from a furnace.

Fuse
(a) An overcurrent protective device with a circuit-opening fusible part that is heated and served by the passage of overcurrent through it. (b) A protection device in an electrical circuit.

Fused Coating
A metallic coating (usually tin or solder alloy) that has been melted and solidified, forming a metallurgical bond to the basis metal.

Fusing
The melting of a metallic coating (usually electrodeposited), followed by solidification.

Fusion
The process whereby the nuclei of light elements, especially those of the isotopes of hydrogen (deuterium and tritium), combine to form the nucleus of a heavier element with the release of substantial amounts of energy.

Fusion Splice
A spliced accomplished by the application of localized heat sufficient to fuse or melt the ends of two lengths of optical fiber, forming a continuous, single fiber.

Fuze
An arming device for a munition.

G

G
Also g. An acceleration equal to the acceleration of gravity, 980.665 centimeters per second per second (approximately 32.2 feet per second per second at sea level), used as a unit of stress measurement of bodies that are being influenced by acceleration.

Gage (Gauge)
(1) An instrument that permits a pass or fail determination of a dimensional characteristic without knowledge of the actual dimension. (2) A device for inspecting/evaluating a limit or size of a specified product dimension.

Gage Accuracy
The difference between the observed average of measurements and the true average of the same parts using precision instruments.

Gage Length
The original length of that portion of the specimen over which strain or change of length is determined.

Gage Linearity
The difference in the accuracy values through the expected operating range of the gage.

Gage Repeatability (Precision)
The variation in measurements obtained with one gage when used several times by one operator while measuring the identical characteristic on the same part.

Gage Reproducibility
The variation in the average of the measurements made by different operators using the same gage while measuring the identical characteristic on the same part.

Gage Stability
The difference in the average of no less than two sets of measurements obtained with a gage on the same parts as a result of time.

Gage System Error
The combination of gage accuracy, repeatability, reproducibility, stability, and linearity.

Gain
The ratio of output current voltage or power to input current, voltage, or power respectively. *Note 1*: Gain is usually expressed in decibels. *Note 2*: Differences in power levels between points in a system may be expressed as gain. *Note 3*: Gain may be expressed as a positive or negative quantity; when a negative quantity, it is usually referred to as *loss*.

Galvanic Cell
A corrosion cell consisting of two dissimilar metals in contact, or in a narrower context consisting of adjacent anodic and cathodic sites: The former definition is on a macro-scale, the latter on a micro scale.

R. Hanifan, *Concise Dictionary of Engineering: A Guide to the Language of Engineering*, DOI 10.1007/978-3-319-07839-7_7, © Springer International Publishing Switzerland 2014

Galvanic Corrosion

Galvanic corrosion manifests itself in the accelerated corrosion caused to the more active metal (anode) of a dissimilar metal couple in an electrolyte solution or medium, and decreased corrosive effects on the less active metal (cathode), as compared to the corrosion of the individual metals, when not connected, in the same electrolyte environment.

Galvanic Series

A galvanic series is a listing of metals and alloys based on their order and tendency to corrode independently, in a particular electrolyte solution or other environment. This tendency for dissolution or corrosion is related to the electrical potential of the metal in conductive medium. Galvanic corrosion is inherently affected by the relative position of the galvanic series of the metals constituting the couple. Metals closely positioned in the series will have electrical potentials nearer one another, whereas with greater divergence in position, greater differences in potential will prevail. Galvanic effects, i.e., corrosion of the anode will in the former condition be minimal; the latter condition will exhibit more significant corrosive effects.

Galvanostat

An electronic device that maintains a constant current flow at an electrode surface.

Gamma

(1) Unit of magnetic intensity. (2) Definite numerical indication of the degree of contrasts in a photograph, facsimile reproduction, or received television picture.

Gas Carbon-Arc Welding

An arc-welding process in which fusion is obtained by heating with an electric arc between a single carbon electrode and the work. Shielding is obtained from a gas or gas mixture (which may contain an inert gas).

Gas Holes

Appears as round or elongated, smooth-edged dark spots on radiograph occurring individually, in clusters, or distributed throughout the casting. They are generally caused by trapped air or mold gases.

Gas Metal-Arc Welding

An arc-welding process in which fusion is obtained by heating with an electric arc between a filler metal electrode (consumable) and the work. Shielding is obtained from a gas (which may contain an inert gas) or a mixture of a gas and a flux. The process is also called *metal inert-gas* (*MIG*) welding.

Gas Pocket

A weld cavity caused by entrapped gas.

Gas Porosity

Minute voids usually distributed through the entire casting. Represented by round or elongated dark spots on radiograph. They are generally caused by trapped air or mold gases rejected during solidification.

Gas Tungsten-Arc Welding

An arc-welding process in which fusion is obtained by heating with an electric arc between a single tungsten (nonconsumable) electrode and the work. Shielding is obtained from an inert gas or gas mixture. Pressure and filler metal may or may not be used. The process is also known as *tungsten inert-gas* (*TIG*) *welding*.

Gate
(1) A channel in a mold used for introducing molten material to the mold cavity. (2) The end of the runner where the molten metal enters the mold. Sometimes this term is applied to the entire assembly of connected channels, feeders, etc., of the metal that fills them. (3) A circuit such as found in electronic computers, with an output and a multiplicity of inputs so that the output is energized only when certain input conditions are met; a circuit in which a signal switches another signal on or off.

Gel
The initially jelly-like solid phase that develops during formation of resin from a liquid. Also, a semisolid system consisting of a network of solid aggregates in which liquid is held.

Gel Coat
A quick-setting resin used in molding processes to provide an improved surface for composites; it is the first resin applied to the mold after the mold-release agent.

Gel Time
The time, expressed in seconds, required for a resin to change its physical state from a solid through a liquid to a solid again due to the action of thermal input.

General License
An unrestricted license or exemption authorized in export-control regulations to export particular technical data, or other items, without obtaining a specific written authority from the Government.

General Specification
A specification prepared in the six-section format, which covers requirements and test procedures that are common to a group of parts, materials, or equipments and is used with specification sheets.

Geometric Tolerance
The general term applied to the category of tolerances used to control form, profile, orientation, location, and runout.

Ghost
A secondary image or signal resulting from echo, envelope delay distortion, or multipath reception.

Gimbal
A device containing two mutually perpendicular and intersecting axis of motion, providing free angular movement in two directions for the mounting of a missile engine. In a gyro, the support providing the spin axis with some degree of freedom.

Glass, Baryta
A type of optical glass containing lead and barium for increasing the refractive index while maintaining a relatively low dispersion.

Glass, Flint
In optics, a glass to which lead and/or other elements are added to produce a generally high refractive index (1.6–1.9) and a low Abbe (v) number (29–51).

Glass, Jena
An improved type of optical glass first made in Jena (Germany) in 1882 by Ernst Abbe and Otto Schott and subsequently used in the photographic lenses and optical instruments produced by Carl Zeiss.

Glass, Opal
A highly diffusing glass having a nearly white, milky, or gray appearance. The diffusing properties are an inherent, internal characteristic of the glass.

Glass Optical
A glass that, during manufacture, is carefully controlled with respect to composition, melting, heat treatment, and other processing so that its optical characteristics (e.g., index of refraction, dispersion, transmittance, spectral transmittance, freedom from birefringence, permanence, etc.) have the values required for the optical application for which it is to be used.

Glazing
The process of inserting lenses into a frame.

Golden Board
A fault-free circuit board used for comparison purposes during testing to identify failed boards. Also referred to as *known good board*.

GO/NO-GO Test
A set of terms (in colloquial usage) referring to the condition or state of operability of a unit that can only have two parameters: (a) GO, functioning properly, or (b) NO-GO, not functioning properly.

Government Bill of Lading (GBL)
A Government document used to procure transportation and related services from commercial carriers.

Government Design Activity (GDA)
The Government agency responsible or scheduled to become responsible for configuration management and design requirements of a configuration item.

Government Furnished Equipment (GFE)
Equipment furnished by the Government that is designed into or will otherwise become a part (or subsystem) of the total system being acquired.

Government Procurement Quality Assurance (PQA)
The function by which the Government determines whether a contractor has fulfilled his contract obligations pertaining to quality and quantity. This function is related to and generally precedes the act of acceptance.

Grade
This term usually implies differences in quality and is usually designed by capital letters; thus, "Grade A," "Grade B," etc.

Grain Boundary
A narrow zone in a metal corresponding to the transition from one crystallographic orientation to another, thus separating one grain from another: the atoms in each grain are arranged in an orderly pattern.

Grain Directiongrain Boundary
The predominant or orientation of the fibrous crystalline structural units of wrought metals.

Grain Flow (Flow Lines)
The directional elongation in the grain structure of the material, and its nonhomogenous constituents, resulting from the forging process. Grain flow follows the direction of working during forging and is usually revealed by polishing and etching sections of the forging.

Graphics
The art or science of conveying information through the use of graphs, letters, lines, drawings, pictures, etc. *Note*: facsimile is one technology for electrically transporting intelligence in graphics form from one point to another.

Graphics Standard
A technical standard describing the digital exchange format of graphics data.

Graphitic Corrosion
A type of attack in gray cast irons in which the metallic constituents are converted to corrosion products, leaving the graphite flakes (cathodes) intact (also called *graphitization*).

Greenwater Loading
Mechanical loading due to wave slap.

Greige
Fabric that has received no finish.

Grid
(1) An orthogonal network of two sets of parallel equidistant lines used for locating points on a printed board. (2) A theoretically exact, two-dimensional networks consisting of a set of equally spaced lines, superimposed upon another set of equally spaced parallel lines, such that the lines on one set are perpendicular to the lines of the other, thereby forming square areas. The intersections of the lines provide the basis for a theoretically exact increment location system.

Grinding and Polishing
A small-scale adaptation of the lens maker's technique of shaping a piece of glass by rubbing it against another object with the intervention of various sizes of abrasive grains mixed with a liquid.

Grinding Cracks
Thermal cracks due to local overheating of the surface being ground, generally caused by lack of, or poor, coolant, a dull wheel or one of improper grain, too rapid a feed, or too heavy a cut. Also called *grinding checks*, because they often appear as a checkered network

Ground
The electrical connection to earth through an earth electrode subsystem. This connection is extended throughout the facility via the facility ground system consisting of the signal reference subsystem, the fault protection subsystem, and the lightning protection subsystem.

Grounding
The bonding of an equipment case, frame, or chassis to an object or vehicle structure to ensure a common potential. The connection of an electric circuit or equipment to earth or to some conducting body of relatively large extent that serves in place of earth.

Ground Plane
A conductor layer, or portion of a conductor layer (usually a continuous sheet of metal with suitable ground plane clearances), used as a common reference point for circuit returns, shielding, or heat sinking.

Ground Plane Clearance
The etched portion of a ground plane around a plated-through or non-plated-through hole that isolates the plane from the hole.

Ground, Single-Point
A scheme of circuit/shield grounding in which each circuit/shield has only one physical connection to ground, ideally at the same point for a given subsystem. This technique prevents return currents from flowing in the structure.

Ground Support Equipment
All equipment (implements, tools, test equipment, devices, mobile or fixed, and so forth) required on the ground to make an aerospace system (aircraft, missile, and so forth) operational in its intended environment.

Ground Support Systems (GSS)
Infrastructure and equipment (portable or fixed) that provides functional or physical support to GSE and does not directly interface with flight hardware, although it may supply commodities, power, or data that eventually reaches the flight hardware after being conditioned or controlled by GSE.

Group
(1) A collection of units, assemblies, or subassemblies that is a subdivision of a set or system, but that is not capable of performing a complete operational function. (2) A collection of units, assemblies, or subassemblies that is not capable of performing a complete operational function. A group may be a subdivision of a set or may be designed to be added to or used in conjunction with a set to extend the function or the utility of the set.

GSI
Formerly known as EAN.UCC, the Uniform Code Council and EAN International have been restructured resulting in a name change to GSI for the combined organization for the establishment of product coding standardization and issuance of unique company prefix codes.

Guarding
A means of connecting an input signal so as to prevent any common-mode signal from causing current to flow in the input; thus, differences of source impedance do not cause conversion of the common-mode signal into a differential signal.

Guided Probe
An electrical probe to measure status of internal nodes on a circuit board to perform fault diagnosis. The operator is "guided" by a software algorithm that uses the results of the previous measurement to determine the next note to be measured.

Guided Probe System
A fault-isolating technique in which the test program causes the ATE display to indicate where the test performer should affix the ATE's diagnostic probe on the UUT. The test program then analyzes the signal sensed by the probe and causes the ATE display to indicate where next to affix the probe. This process continues until the fault has been isolated to the best of the test program's capability.

H

Half-Duplex
The property of a channel that allows communications in either direction, but only in one direction at a time.

Halftone
Any photomechanical printing surface, or the impression there from, in which detail and tone values are represented by a series of evenly spaced dots in varying size and shape, varying in direction proportion to the intensity of tones they represent.

Halogenated Solvent
A halogenated solvent is any solvent whose chemical structure contains bromine, chlorine, fluorine, or iodine. Most halogenated solvents are non-flammable and many are very toxic.

Haloing
Mechanically induced fracturing or delaminating on or below the surface of the base material; it is usually exhibited by a light area around holes, other machined areas, or both.

Hand Lay-Up
A process in which components are applied either to a mold or a working surface, and the successive plies are built up and worked by hand.

Handshake
(1) An interlocked exchange of signals between a master and a slave, controlling the transfer of data. (2) A hardware or software sequence of events requiring mutual consent of conditions prior to change.

Hangfire
The delayed ignition of a rocket propellant or igniter.

Hard Conversion
The process of changing inch-pound measurement units to nonequivalent metric units, which necessitates physical conversion changes outside those permitted by established measurement tolerances. *Note*: although the term *hard conversion* is in general use, it is technically incorrect when applied to specific items because no "conversion" takes place. Instead, a new metric item requiring new item identification is created to replace the customary item. The new item is often referred to as being in *hard metric*.

Hard Copy
In computer graphics, a copy of display image on a data medium such as paper or polyester film.

Hardened Structure
A structure that has been strengthened sufficiently to withstand nuclear weapon effects.

Hardening
Increasing the hardness by suitable treatment, usually involving heating and cooling.

Hardness Critical Item (HCI)
An item of hardware or software that satisfies one or more of the following conditions:

- Functionally required hardware whose response to the specified nuclear environments could cause degradation in systems survivability unless additional provisions for hardness are included in the item specification, design, manufacture, item selection process, provisioning, configuration control, etc.
- Functionally required hardware or software that inherently provides protection for the system or any of its elements against the specified nuclear environments and that, if modified, removed, or replaced by an alternate design could cause a degradation in system survivability
  * Hardness dedicated hardware or software included in the system solely to achieve system nuclear survivability requirements
  * Hardware items (at the level of application) to which a Hardness Critical Process is applied, and/or
  * A subassembly or higher level of assembly that contains one or more HCIs

Hardness Critical Process (HCP)
Any fabrication, manufacturing, assembly, installation, maintenance, repair, or other process or procedure that implements a hardness design feature and satisfies system hardness requirements.

Hard Points
On a missile's exterior surface, structurally strengthened areas suitable for support during handling.

Hardware
Items made of material, such as weapons, aircraft, ships, tools, computers, vehicles, and their components (mechanical, electrical, electronic, hydraulic, and pneumatic). Computer software and technical documentation are excluded.

Hardware Configuration (HWCI)
A configuration item that is hardware.

Hardwire
To make permanent connections (such as soldered and wirewrapped connections) between circuits; as contrasted with quick-disconnect connections (such as plug-in, threaded or twist-lock connections).

Hard Wiring
Hard wiring is electrical wiring that interconnects two or more components or assemblies into an assembly that is inseparable without the use of special tools and techniques.

Harmonic
Of a sinusoidal wave, an integral multiple of the frequency of the wave. *Note*: the first harmonic is the fundamental frequency itself; the second harmonic is twice the frequency of the fundamental; the third, three times the frequency of the fundamental; and so on.

Harmonic Distortion
The presence of frequencies at the output of a device that are not present at the input, caused by nonlinearities within the device and harmonically related to a single frequency applied to the input of the device.

Harness
(1) An assembly of wires, cables, or both, arranged so it may be installed or removed as a unit in the same electronic or electrical equipment. (2) An assembly of any number of wires, cables, and/or groups and their terminations, designed and fabricated so as to allow for installation and removal as a unit. A harness may be an open or protected harness.

Hazard
(1) A condition that is prerequisite to a mishap. (2) A potential source of harm.

Hazardous Atmosphere
The term *hazardous atmosphere* is a vapor or gas mixture with air that, under normal conditions of temperature and pressure, can be ignited by a spark or flame to produce a sustained, self-propagating, combustion wave, or that is toxic or asphyxiant.

Hazardous Event Probability
The likelihood, expressed in quantitative or qualitative terms, that a hazardous event will occur.

Hazardous Materials
An item of supply consisting of materiel that because of its quantity, concentration, or physical, chemical, or infectious characteristics, may either cause or significantly contribute to an increase in mortality or an increase in serious, irreversible, or incapacitating reversible illness; or pose a substantial present or potential hazard to human health or the environment when improperly treated, stored, transported, disposed of, or otherwise managed. (This includes all items listed as hazardous in Titles 29, 40, 49 CFR and other applicable modal regulations effective at the time of shipment.)

Hazardous Waste
Those wastes that require special handling to avoid illness or injury to persons or damage to property.

Hazard Probability
The aggregate probability of occurrence of the individual hazardous events that create a specific hazard.

Hazard Severity
An assessment of the worst credible mishap that could be caused by a specific hazard.

Header
Control information, appended to every burst, that defines the type and format of the burst, and the network identification numbers of the source and designation.

Header Extension
An extension to the header that is necessary to support certain burst formats, including acknowledgment, multicast, and multicast probe bursts.

Header Kernel
The core of the header, contained in all headers.

Heads-Up Display (HUD)
A positional/command display typically projected on the primary windscreen of an aircraft cockpit to provide the pilot with operational information without having to look down at the cockpit instrument panel displays.

Heat Affected Zone
An area adjacent to a weld where heat has caused microstructural changes that affect the corrosion behavior of the metal.

Heat Exchanger
An air-to-air or liquid-to-air finned duct arrangement that is used to transfer dissipated heat from a hot recirculating fluid to the cooling fluid by conduction through the finned surfaces.

Heat Sinking Plane
A continuous sheet of metal on or in a printed board that functions to dissipate heat away from heat-sensitive components.

Heat-Exchanger Tube
A tube for use in apparatus in which fluid inside the tube will be heated or cooled by fluid outside the tube. The form usually is not applied to coil tube or tubes for use in refrigerators or radiators. (Note This product is typically seamless drawn tube.)

Heat Treatment
Heating and cooling a solid metal or alloy in such a way as to obtain desired conditions or properties.

Helix, Direction of
When a spring is viewed from one end, the direction of helix is right hand when the coil recedes in a clockwise direction and left hand when it recedes in a counterclockwise direction. A right-hand helix follows the same direction as the threads on a standard screw. A left-hand helix is most popular for compression and extension springs.

Henry
Unit of inductance in which the inducted electromotive force of one volt is produced by the inducing current changing at the rate of one ampere per second.

Hermetic Seal
(1) An impervious seal made by the fusion of metals or ceramic materials (as by brazing, soldering, welding, or fusing glass or ceramic) that prevents the passage of gas or moisture. (2) The process by which an item is totally enclosed by a suitable metal structure or case by fusion of metallic or ceramic materials. This includes the fusion of metals by welding, brazing, or soldering; the fusion of ceramic materials under heat or pressure; and the fusion of ceramic materials into a metallic support.

Hertz (Hz)
A unit of frequency representing cycles per second.

Hierarchical Structure
An organization of software modules that consists of a root node. This term can be applied to data as well as program. This structure is also known as *tree structure without cycles*.

Hierarchy
(1) A series of successive tasks or routines in a graded order. (2) A structure by which objects or classes of objects are ranked according to some subordinating principle or set of principles. One common representation of a hierarchy is the directed tree-graph, in which the root node heads the hierarchy and all other objects are ranked by order into levels of subordination. If a single subordinating relationship governs, the hierarchy is said to be *unordered*; otherwise, it is *ordered*.

Higher Frequency Ground
The interconnected metallic network intended to serve as a common reference for currents and voltages at frequencies above 30 kHz, and in some cases above 300 kHz. Pulse and digital signals with rise and fall times of less than 1 microsecond are classified as higher frequency signals.

High Order Mode
A propagation path that makes a relatively large angle with respect to the fiber axis.

Hi-Pot
A test designed to determine the highest potential that can be applied to a conductor without breakdown of the insulation.

Histogram
A pictorial way to display data in frequency form. This provides a visual way to evaluate the form of the data.

Hole Density
The quantity of holes in a printed board per unit area.

Hole Pull Strength
The force necessary to rupture a plated-through hole when loaded or pulled in the direction of the axis of the hole.

Hole Void
A void in the metallic deposit of a plated-through hole exposing the base material.

Homogeneous
Descriptive term for a material of uniform composition throughout; a medium that has no internal physical boundaries; a material whose properties are constant at every point, i.e., constant with respect to spatial coordinates (but not necessarily with respect to directional coordinates).

Homogeneous Armor
An armor made from a single material that is consistent throughout in terms of chemical composition, physical properties, and degree of hardness.

Homogeneous Cable
Cable composed of identical insulated conductors.

Hooking
A task performed by a human operator using a display interactive device (e.g., joystick or mouse) to select/designate specific display information (e.g., symbols) for further action or modification.

Hookup Wire
Refers to an insulated conductor, free at both ends and used for chassis wiring and interconnecting wiring.

Horizontal Situation Display (HSD)
An HSD is a display that aids crew members in navigation. Basically, it consists of heading, distance-to-go, bearing-to-destination or some other navigation facility or reference, track, map, course, aircraft position, and steering error. Modes may consist of manual, north-up, track-up, data, test, and off. Selection of map scale factors may also be provided. Navigation

update can be accomplished with the proper computer techniques. The HSD also has the capability of combining symbols with the map information. Symbols may be used for annotation of the projected map, such as check points, various legs of the mission, high-risk areas, ground track deviation, and radar homing and threat warning. Specific modes and formats can be selected for a given mode of operation.

Host Space Vehicle
The space vehicle that contains a payload or provides auxiliary support devices (in the form of electrical power and so forth) to the payload.

Hot Gas Welding
A technique used to join thermoplastic materials (usually sheet). The materials are softened by a jet of hot air from a welding torch and joined together at the softened points. Generally, a thin rod of the same material is used to fill and consolidate the gap.

Hot Mockup
Any assemblage of repair parts, components, modules, or similar items configured to simulate an end item or subsystem for the purpose of testing or checking individual or collective parts, components, modules, or similar items.

Hot Salt Cracking
A type of stress-corrosion cracking of titanium alloys occurring when in contact with chloride salts at temperatures in the range of 288–538°C.

Hot Welded Can
A hot welded can is a cap-sealed component utilizing thermocompression attachment of the cap to the base of the device.

Hover, In Ground Effect
A condition in which a helicopter is motionless with respect to the ground, and the rotor is operating at one rotor diameter height or less above ground level.

Human Engineering
The area of human factors that applies scientific knowledge to the design of items to achieve effective operation, maintenance, and man/machine integration.

Human Engineering Design Criteria
The summation of available knowledge that defines the nature and limits of human capabilities as they relate to the checkout, operation, maintenance, or control of systems or equipment, and that may be applied during engineering design to achieve optimum compatibility between equipment and human performance.

Human Factors
Human characteristics relative to complex systems and the development and application of principles and procedures for accomplishing optimum man–machine integration and utilization. The term is used in a broad sense to cover all biomedical and psychosocial considerations pertaining to man in the system.

Human-Readable Information (HRI)
Information intended to be conveyed to a person. HRI in lieu of machine-readable information is commonly referred to as text. HRI applications in association with a linear bar code or two-dimensional (2D) symbol are identified as (a) human-readable interpretation, (b) human translation, (c) data area titles, and (d) free text.

Human-Readable Interpretation
Human readable information provided adjacent to a machine-readable medium representing the encoded data within the medium.

Human Translation
Human-readable information provided within proximity of the machine readable medium, representing portions of the information encoded and data field descriptions not encoded in the symbols.

Hybrid
(1) A composite laminate composed of laminae of two or more composite material systems. Or, a combination of two or more different fibers, such as carbon and glass or carbon and aramid, into a structure (tapes, fabrics, and other forms may be combined). (2) A device, circuit, apparatus, or system made up of two or more subassemblies, often employing different technologies, not heretofore combined, to satisfy a given requirement. *Note*: examples include (a) an electronic circuit having both vacuum tubes and transistors, (b) a mixture of thin-film and discrete integrated circuits, and (c) a computer that has both analog and digital capability. (3) A transformer or combination of transformers, or resistive network affording paths to three branches, circuits A, B, and C, so arranged that A can send to C, B can receive from C, but A and B are effectively isolated.

Hybrid Circuit
A circuit possessing both digital and analog signals.

Hybrid Item
An item designed and produced using both metric and inch-pound units even though it may be described by only one system of units in standardization documents.

Hybrid Specification
A specification in which some requirements are given in rounded, rational metric units, and other requirements are given in rounded, rational inch-pound units. Hybrid specifications are often required for use in new design where existing usable components must interface in a metric system.

Hydrogen Embrittlement
A reduction of the load-carrying capability by entrance of hydrogen into the metal (e.g., during pickling or cathodic polarization).

Hydrogen-Stress Corrosion
A premature failure of a metal resulting from the combined action of tensile stresses and the penetration of hydrogen into the metal.

Hypersonic
Having a high velocity. Usually applied to velocities of more than Mach 5.

I

Icicle
- see SOLDER PROJECTION.

Identification Cross Reference Drawing
An administrative-type drawing that assigns unique identifiers that are compatible with automated data processing systems and item identifications specifications, and provides a cross reference to the original incompatible identifier.

Identification Plate
A plate used to identify equipment with nomenclature type designation, manufacturer, part number, etc. It does not normally contain instructional, caution, shipping or maintenance information.

Identification Test
A test used to verify that the item to be tested is the correct unit.

Image, Aspect of
A term denoting the orientation of the image, such as normal, canted, inverted, or reverted.

Image Card
A completely processed aperture, camera, or copy card. A completely processed card must have eye-readable header information and be keypunched, and its aperture must contain a micro-image. So, an image card is created when (a) a frame of developed silver halide camera microfilm is inserted in the aperture of an unprocessed aperture card, (b) exposing and developing the camera microfilm in an unprocessed camera card, or (c) using an unprocessed copy card to reproduce an existing image card (i.e., a processed camera, copy, or aperture card).

Immersion Plating (Galvanic Displacement)
The chemical deposition of a thin metallic coating over certain basic metals by a partial dis placement of the basic metal.

Immunity
A state of resistance to corrosion or anodic dissolution caused by the fact that the electrode potential of the surface in question is below the equilibrium potential of anodic dissolution.

Impact Resistance
Relative susceptibility of plastics to fracture by shock, e.g., as indicated by the energy expended by a standard pendulum-type impact machine in breaking a standard specimen in one blow.

Impact Strength
(1) The ability of a material to withstand shock loading. (2) The work done in fracturing, under shock loading, a specified test specimen in a specified manner.

Impedance
The total passive opposition offered to the flow of electric current. It consists of a combination of resistance, inductive reactance, and capacitive reactance. *Note*: impedance is usually a function of frequency.

R. Hanifan, *Concise Dictionary of Engineering: A Guide to the Language of Engineering*,
DOI 10.1007/978-3-319-07839-7_9, © Springer International Publishing Switzerland 2014

Imperfect Thread
- see INCOMPLETE THREAD.

Impingement Attack
A form of localized corrosion-erosion caused by turbulence or impinging flow.

Impulse
A surge of electrical energy, usually of short duration, of a nonrepetitive nature.

Inactive Time
The period of time when an item is available, but it is neither needed nor operated for its intended use.

Inch-Pound Specification
Inch-pound specifications have requirements given in rounded, rational, inch-pound units, usually as a result of being originally developed in inch-pound. The magnitudes are meaningful and practical. Inch-pound specifications should include those with rounded rational, inch-pound units only. *Note*: There have been instances in which magnitudes expressed in metric units as a result of mathematical conversion from rounded rational, inch-pound units are given first (preferred units) with the rounded, rational, inch-pound units given in parenthesis or in a nonpreferred position. These specifications are inch-pound documents. Inch -pound specifications are developed for items to interface or operate with other inch-pound items.

Inch-Pound Units
The customary system formerly and currently used in the United States (foot, inch, pound, BTU, horsepower, degree Fahrenheit, etc.).

Inclusion
(1) A term used to denote the presence, within the body of the glass, of extraneous or foreign material. (2) Inclusions are particles of foreign material such as sand or slag that are embedded in the cast metal. (3) A foreign particle, metallic or nonmetallic, in a conductive layer, plating, or base material.

Incomplete Fusion
A discontinuity caused by localized temperatures that are insufficient to allow fusion of the molten weld metal and nonmolten weld or base metals.

Incomplete Thread
A threaded profile having either crests or roots, or both crests and roots not fully formed, resulting from the intersection with the cylindrical or end surface of the work or the vanish cone. It may occur at either end of the thread.

Independency (I)
A symbol used in geometric dimensioning and tolerancing that indicates that perfect form of a feature of size at MMC or at LMC is not required.

Independe'nt Failure
A failure that occurs without being caused by or related to the failure of associated items, distinguished from *dependent failure*.

Indexing Holes
Indexing holes are holes placed in a printed-board blank to enable accurate positioning for processing. (Indexing holes may or may not be on the finished board.)

Indexing Notch
An indexing notch is a notch placed in the edge of a printed board blank to enable accurate positioning for processing. (Indexing notches may or may not be on the finished board.)

Index List (IL)
A tabulation of data lists and subordinate index lists pertaining to the item to which the index applies.

Indicator-Light Housing
An electrical device that accommodates a lamp, both electrically and mechanically, and provides for the use of an indicator-light lens.

Inductance
The property of a circuit or circuit element that opposes a change in current flow. Inductance thus causes current changes to lag behind voltage changes. Inductance is measured in henrys.

Induction Soldering (IS)
A soldering process in which the heat required is obtained from the resistance of the work to induced electric current.

Industry Standard
An industry-developed document that establishes voluntary specifications for products, practices, or operations.

Inert Gas
Gases that are completely unreactive.

Infinity
In the optical industry, a term used to denote a distance sufficiently great that light rays emitted from a body at the distance are practically parallel.

Information Plate
Plates used to present information other than identification, such as warning, ratings, wiring connections, diagrams, charts, operating or maintenance instructions, etc.

Infrared
The visible electromagnetic radiation beyond the red end of the visible spectrum. The wavelengths range from 768 millimicrons to the region of 30 or 40 microns. Heat is radiated in the infrared region.

Infrared Brazing (IRB)
A brazing process in which the heat required is furnished by infrared radiation.

Inherent Testability
A testability measure that is dependent only upon hardware design and is independent of test stimulus and response data.

Inherent Value
A measure of maintainability that includes only the effects of item design and application and assumes an ideal operation and support environment.

Inhibitor
A chemical substance or combination of substances that, when present in the proper concentration and forms in the environment, prevents or reduces corrosion by a physical, physical-chemical, or chemical action.

Initialize
To place an item into a known state.

Initial Graphics Exchange Specification (IGES)
A neutral file format for the representation and transfer of product definition data among CAD/CAM systems and application programs.

Initial Point
The point relative to which all pels are positioned within a page. It does not necessarily lie at a corner of a tile in the tile grid.

Injection Blow Molding
A blow-molding process in which the parison to be blown is formed by injection molding.

Injection Molding
A molding procedure whereby a heat-softened plastic material is forced from a cylinder into a relatively cool cavity that gives the article the desired shape.

Input
(1) Pertaining to a point in a device, process, or channel at which it accepts data. (2) An input state or a sequence of states.

Input Data
(1) Data being received, or to be received, by a device or a computer program. (2) Data to be processed.

Input/Output Device
A device that introduces data into or extracts data from a system.

Input–Output Channel
For a computer, a device that handles the transfer of data between internal memory and peripheral equipment.

Input Protection
For analog input channels, protection against overvoltages that may be applied between any two input connectors or between any input connector and ground.

Inseparable Assembly
(1) Multiple pieces joined by such processes as bonding, riveting, brazing, welding, etc. that are not normally subjected to disassembly without destruction or impairment of the designed use. (2) Same as part.

Insertion Loss
The reduction in power that takes place at the load on insertion of a network between the source and the load. It is generally expressed as a ratio in decibels.

Inspection
Inspection is the process of measuring, examining, testing, or otherwise comparing a unit of product with the requirements.

Inspection by Attributes
Inspection whereby either the unit or product is classified simply as defective or nondefective, or the number of defects in the unit of product is counted, with respect to a given requirement or set of requirements.

Inspection by Variables
Inspection wherein certain quality characteristics of sample are evaluated with respect to a continuous numerical scale and expressed as precise points along this scale. Variables inspection records the degree of conformance or nonconformance of the unit with specified requirements for the quality characteristics involved.

Inspection, Cyclical
A system whereby supplies and equipment in storage are subjected to, but not limited to, periodic scheduled and special inspection and continuous action to assure that material is maintained in a ready-for-issue condition.

Inspection, In-Process
Inspection performed during the manufacturing or repair cycle in an effort to prevent defectives from occurring and to inspect the characteristics and attributes that are not capable of being inspected at final inspection.

Inspection Level
An indication of the relative sample size for a given amount of product.

Inspection Lot
A collection of units of product bearing identification and treated as a unique entity form which a sample is to be drawn and inspected to determine conformance with the acceptability criteria.

Inspection, Original
First inspection of a particular quantity of product, as distinguished from the inspection of product that has been resubmitted after prior rejection.

Inspection Overlay
A positive or negative transparency made from the production master and used as an inspection aid.

Inspection, Quality Conformance
All examinations and tests performed on items or services for the purpose of determining conformance with specified requirements.

Inspection Record
Recorded data concerning the results of inspection action.

Inspection, Reduced
Inspection under a sampling plan using the same quality level as for normal inspection, but requiring a smaller sample for inspection.

Inspection System Requirement
A requirement to establish and maintain an inspection system in accordance with Government specification Mil-I-45208. The requirement is referenced in contracts when technical requirements are, e.g., to require control of quality by in-process inspection as well as final, end item inspection.

Inspection, Tightened
Inspection under a sampling plan using the same quality level as for normal inspection, but requiring more stringent acceptance criteria.

Installation (Complete Equipment)
An installation (complete equipment) is defined as a combination of assemblies, accessories, and detail parts required to make one complete operating equipment. An installation comprises a group of permanently installed parts and a group of removable assemblies.

Installation Drawing
An installation drawing provides information for properly positioning and installing items relative to their supporting structure and adjacent items, as applicable.

Insulation Resistance
The electrical resistance of the insulating material (determined under specified conditions) between any pair of contacts, conductors, or grounding devices in various combinations.

Interference Fit
A fit between mating assembled parts that always provides an interference.

Integral Composite Structure
Composite structure in which several structural elements, which would conventionally be assembled by bonding or with mechanical fasteners after separate fabrication, are instead laid up and cured as a single, complex, continuous structure; e.g., spars, ribs, and one stiffened cover of a wing box fabricated as a single integral part. The term is sometimes applied more loosely to any composite structure not assembled by mechanical fasteners.

Integral Parts List
A parts list prepared and revised as part of an engineering drawing.

Integrated Logistic Support (ILS)
A disciplined, unified, and iterative approach to the management and technical activities necessary to (a) integrate support considerations into system and equipment design; (b) develop support requirements that are related consistently to readiness objectives, to design, and to each other; (c) acquire the required support; and (d) provide the required support during the operational phase at minimum cost.

Integrated Standard
A technical standard describing the exchange format of digital data that integrates text, graphics, alphanumeric, and other types of data in a single (compound) file.

Intensity Modulation
By controlling current to the light source, the intensity of the light is made to vary over a continuous range more or less in proportion to the applied signal.

Intent
May: This word is understood to be permissive. *Shall*: This word is understood to be mandatory. *Should*: This word is understood to be advisory.

Interactive Graphics
The use of a display terminal in a conversational mode or interactive mode.

Interactive Mode
- see CONVERSATIONAL MODE.

Interchangeability
Refers to features that affect form, fit, or function of an item in such a way as to enable common use or replacement with like items without selective assembly or modification.

Interchangeable Item
(1) One that possesses such functional and physical characteristics as to be equivalent in performance to another item of similar or identical purposes; and is capable of being exchanged for the other item without selection for fit or performance, and without alteration of the items themselves or of adjoining items, except for adjustment. (2) Items are interchangeable when the items possess functional and physical characteristics that are equivalent in performance and durability and are capable of being exchanged one for the other without alteration of the items themselves or of adjoining items except adjustment, and without selection for fit or performance. (3) One that (a) possesses such functional and physical characteristics as to be equivalent in performance, reliability, and maintainability to another item of similar or identical purposes and (b) is capable of being exchanged for the item without selection for fit or performance and without alteration of the items themselves or of adjoining items, except for adjustment.

Intercharacter GAP
The space between the last element of one character and the first element of the adjacent character of a discrete bar code.

Interconnecting Cable
Two or more insulated conductors contained in a common covering, or one or more insulated conductors with a gross metallic shield outer conductor used to carry electrical current between units.

Interconnecting Wire
Insulated, single-conductor wire used to carry electric current between units.

Interconnecting Wiring
Interconnection wiring consists of wires, cables, groups, or harnesses used to connect complete units of electrical or electronic systems.

Interconnection Diagram
A form of a Connection or Wiring Diagram that shows only external connection between unit assemblies or equipment. The internal connections of the unit assemblies or equipment are usually omitted.

Interface
The functional and physical characteristics required to exist at a common boundary.

Interface Characteristic
Those characteristics that affect the physical or functional characteristics of co-functioning items. The characteristics are established to allow equipment or systems to be compatible with equipment or systems under the control of different customers, contractors, or design activities. Changes to interface characteristics shall be coordinated with all affected activities.

Interface Control
The process of identifying, documenting, and controlling all functional and physical characteristics relevant to the interfacing of two or more items provided by one or more organizations.

Interface Control Documentation (ICD)
Interface control drawing or other documentation that depicts physical and functional interfaces of related or co-functioning items.

Interference Fit
An interference fit is one having limits of size so prescribed that an interference always results when mating parts are assembled.

Intergranular Corrosion
A type of corrosion attack that occurs preferentially at grain boundaries (also called *intercrystalline corrosion*).

Interim Amendment
An interim amendment is a limited coordination amendment to a coordinated specification required by a single military department or activity to meet a need when time does not permit preparation of a coordinated amendment.

Interior Container
A container that is inside another container. It may be a unit pack or an intermediate container that is placed inside an exterior container or shipping container.

Internal Conductive Interfaces (Printed Wiring Board)
An internal conductive interface is considered to be the junction between the internal layers (copper foil posts or internal layers) and the deposited or plated copper.

Interlayer Connection
An electrical connection between conductive patterns in different layers of a multilayer printed board.

Intermediate Bearing Stress
The bearing stress at the point on the bearing load-deformation curve where the tangent is equal to the bearing stress divided by a designated percentage (usually 4%) of the original hole diameter.

Intermediate Container
A wrap, box, or bundle containing two or more unit packs of identical items.

Intermediate Level Maintenance
Maintenance that is normally the responsibility of, and performed by, designated maintenance activities for direct support of using organizations. Its phases normally consist of (a) calibrating, repairing, or replacing damaged or unserviceable parts, components, or assemblies and (b) modification of material, emergency manufacturing of unavailable parts. Intermediate maintenance is normally accomplished by the using commands in fixed or mobile shops.

Intermediate Pack
A wrap, box, or bundle that contains two or more unit packs of identical items.

Intermittent Weld
A weld wherein the continuity of the weld is broken by recurring unwelded spaces.

Intermodulation
The mixing of two or more signals in a nonlinear element to produce signals at new frequencies that are sums and differences of the input signals or their harmonics. The nonlinear elements may be internal to the system, subsystem, or equipment, or may be some external device.

Internal Oxidation
A formation of isolated particles of corrosion products beneath the surface of a metal, resulting from inward diffusion of oxygen, nitrogen, sulfur, etc. (also known as *subsurface corrosion*).

International Traffic in Arms Regulations (ITAR)
The regulations issued by the Department of State under the AECA and printed at 22 CFR 121–130.

Internet Meme
An internet meme is a concept or idea that spreads "virally" from one person to another via the internet.

Interrupted Diameter
A cylindrical or spherical part whose surfaces are not smooth.

Interstice
A minute space between one thing and another, especially between things closely set or between the parts of a body.

Inverter
(1) In electrical engineering, a device for converting direct current into alternating current.
(2) In computers, a device or circuit that inverts the polarity of a signal or pulse.

Investment Casting
Investment casting involves pouring metal into a mold produced by surrounding (investing) an expendable pattern with a refractory slurry that sets at room temperature. After this, the wax, plastic, or frozen mercury pattern is removed through the use of heat. This procedure is also known as *precision casting* or the *lost wax process*.

Ion
An atom or group of atoms that carries a positive or negative electric charge as a result of having lost or gained one or more electrons.

Irradiation
The exposure of a material to high-energy emissions. In insulations, for the purpose of favorably altering the molecular structure.

Isokeraunic (or Isoceraunic)
Showing equal frequency of thunderstorms.

Isolation
Physical and electrical arrangement of the parts of an equipment, system, or facility to prevent uncontrolled electrical contact within or between the parts.

Isometric Projection
An isometric projection is an axonometric projection in which the three axes of the object make equal angles with the plane of projection. Taken two at a time, the three axes make three equal angles of 120° on the drawing.

Isotropic
Having uniform properties in all directions. The measured properties of an isotropic material are independent of the axis of testing.

Issuing Agency Code (IAC)
The IAC represents the registration authority that issued the enterprise identifier.

Item
(1) A nonspecific term used to denote any unit or product including materials, parts, assemblies, equipment, accessories, and computer software. (2) A single hardware article or a single unit formed by a grouping of subassemblies, components, or constituent parts. (3) A general term used to denote any unit of product or data, including materials, parts, assemblies, equipment accessories, computer software, or documents that have entity.

Item, Commercial

An item designed and available for purchase on the commercial market and included in the term *item, privately developed*.

Item Description (Nomenclature)

The name and description of an item as it appears in the contract, purchase order, or requisition. The source document for this information is the DD Form 61 (Request for Nomenclature), which contains the exact name and description of an item.

Item, End

An item in the assembled or completed state at which Government procurement takes place.

Item Identification

(1) The combination of the part or identifying number and the original design activity CAGE code. (2) The part or identifying number or descriptive identifier for a specific item. For Government applications, the item identification also includes the CAGE Code of the responsible design activity as a prefix to the part identifying number, separated by a dash. If no CAGE code applies, the manufacturer's name and address are needed on the drawing for complete item identification. (3) The part number, identifying number, or descriptive identifier for a specific item along with the enterprise identifier of the activity that assigned the part number, identifying number, or descriptive identifier.

Item, Minor Hardware

An item susceptible to simple correction or open stock replacement, or an item on which all inspection characteristics are unclassified or minor.

Item Name

A name published in the Federal Cataloging Handbook H6, or that name developed by the requestor in accordance with ASME Y14.100, the portion applicable to drawing titles. Item names used with type designator assignments will be consistent with the policies of the Federal Cataloging Program. Examples of unacceptable item names include abbreviations, acronyms, descriptions of size, frequencies, etc.

Item, Privately Developed

An item completely developed at private expense and offered to the Government as a production article, with Government control of the article's configuration normally limited to its form, fit, and function (includes commercial items).

Item Specification

A type of program-unique specification that describes the form, fit, and function and method for acceptance of parts, components, and other items that are elements of a system.

Item, Standard Stock

An off-the-shelf item identified as an industry standard item or Government standard item.

Item Unique Identification (IUID)

A system establishing unique item identifiers within the Department of Defense by assigning a machine-readable character string or number to a discrete item, which serves to distinguish it from other like and unlike items.

IUID Equivalent

Unique identification methods in commercial use that have been recognized by DoD as IUID equivalents which are the Global Individual Asset Identifier (GIAI), Global Returnable Asset Identifier (GRAI when serialized), Vehicle Identification Number (VIN), and Electronic Serial Number (ESN for cell phones only).

J

Jacket
A covering or casing, specifically a shell around the combustion chamber of a liquid Propellant Rocket, through which some of the propellant is circulated for regenerative cooling.

Jacketed Cable
A bundle of insulated wires encased in a common sheath.

Jammer, Automatic Search
An intercept receiver and jamming transmitting system that automatically searches for and jams enemy signals of specific radiation characteristics.

Jammer, Electronic
The intentional radiation or reradiation of electromagnetic waves with the object of impairing the enemy's use of a specific portion of the electromagnetic spectrum.

Jammer, Repeater
A jammer that radiates a signal on the frequency of the enemy equipment, the reradiated signal being so modified as to cause the enemy equipment to present erroneous data on azimuth, range, and number of targets.

Jamming, Spot
Jamming on a specific or narrow band of frequencies. May be active (transmissions) jamming or passive (window) jamming.

J-Display
Radar display in which the time base is a circle. The target signals appear as radial deflections from the time base.

Jet Penetration
Jet penetration in armor achieved by a high-pressure flow process associated with the functioning of shaped-charged warheads.

Jitter
Short-time instability of a signal. The instability may be in either amplitude or phase, or both. Random departure from regularity of repetition.

Jitter, Time
A measure of the uncertainty of the repetitive position of a time mark. Time-related, abrupt, spurious variations in the duration of any specified, related interval.

Joint Penetration
The minimum depth a groove or flange weld extends from its face into a joint, exclusive of reinforcement.

Joule Effect
The heating effect produced by the flow of current through a resistance. The rate at which heat is produced in an electric circuit having constant resistance is proportional to the square of the current.

R. Hanifan, *Concise Dictionary of Engineering: A Guide to the Language of Engineering*, 147
DOI 10.1007/978-3-319-07839-7_10, © Springer International Publishing Switzerland 2014

Joystick
A stick-type input device used to provide continuous control movement. In computer applications, joysticks are typically used to provide two-axis cursor control on a display screen. Joysticks may be displacement or force-operated. A displacement (or isotonic) joystick moves in the direction it is pushed. Displacement joysticks are usually spring loaded so that they return to their center position. A force-operated (or isometric) joystick has no perceptible movement; its output is a function of the force applied.

Jump
The angular displacement of a projectile or missile from the line of elevation and direction at the time the projectile or missile leaves the tube or missile launcher. Also called *jump angle*.

Jumper
An electrical connection between two points on a printed board added after the intended conductive pattern is formed.

Jumper Wire (Haywire)
A wire used as a jumper or a discrete electrical connection that is part of the original design and is used to bridge portions of the basic conductive pattern formed on a printed board.

Junction
A connection between two or more conductors or two or more sections of transmission line, or a contact between two dissimilar metals or materials, as in a rectifier or thermocouple.

Justify
(1) To shift the contents of a register or a field so that the significant character at the specified end of the data is at a particular position. (2) To align text horizontally or vertically so that the first and last line of the text are aligned with their corresponding margins. The last line of a paragraph is often not justified.

K

R. Hanifan, *Concise Dictionary of Engineering: A Guide to the Language of Engineering,* 149
DOI 10.1007/978-3-319-07839-7_11, © Springer International Publishing Switzerland 2014

K-Display
Radar display in which a target appears as a pair of vertical deflections. Difference in pulse height indicates direction and magnitude of antenna azimuth pointing error.

Key
A device designed to assure that the coupling of two components can occur in only one position.

Keying Slot
A slot in a printed board that permits the printed board to be plugged into its mating receptacle but prevents it from being plugged into any other receptacle.

Keyway
A general term covering both keying slots and polarizing slots.

Kicker
A strip of wood nailed to the floor to retrain other dunnage bracing.

Kill Probability
The chance that a target will be destroyed by a given operation. The likelihood of producing the desired kill under the conditions specified. Kill probability is a function of guidance accuracy and the radius of warhead action.

Kilohertz (kHz)
A unit of frequency equal to one thousand cycles per second.

Kilowatt
A unit of power equal to one thousand watts.

Kit Drawing
A kit drawing identifies an item or group of items with instructions for their use. The kit does not necessarily define a complete functional assembly. A kit drawing establishes item identification for the kit, not for the items in the kit.

Knife Line Attack
A form of weld decay sometimes observed on stabilized stainless steel in which the zone of attack is very narrow and close to the weld.

Known Good Board
A fault-free circuit board.

Knuckle Area
The area of transition between sections of different geometry in a filament wound part.

Knurl
Tool with teeth on its periphery used to produce an imprint of the teeth on the cylindrical surface of the work.

L

Label
An item marked with the identification information of another item and affixed to that other item. A label may be of any material similar to or different from that of the item to which it is affixed. A label may be made of a metallic or nonmetallic material. Labels may be affixed to the identified item by any appropriate means. Labels are often referred to as *plates* (i.e., data plate, name plate, ID plate, etc.). However, label material and methods of marking and affixing have no bearing on this distinction.

Laboratory Reference Standards
Standards that are used to assign and check the values of laboratory secondary standards.

Laboratory Secondary Standards
Standards that are used in the routine calibration tasks of the laboratory.

Laboratory Working Standards
Those standards that are used for the ordinary calibration work of the standardizing laboratory. *Note:* laboratory working standards are calibrated by comparison with secondary standards of that laboratory.

Lading
The load or cargo being shipped.

Laminate
A product made by bonding together two or more layers of material.

Laminate Orientation
The configuration of a cross-plied composite laminate with regard to the angles of cross plying, the number of laminae at each angle, and the exact sequence of the lamina lay-up.

Laminate Thickness
Thickness of the metal-clad base material, single or double-sided, prior to any subsequent processing.

Lampholder
An electrical device that accommodates a lamp, both electrically and mechanically, but does not provide for the use of an indicator-light lens.

Land
A portion of a conductive pattern usually, but not exclusively, used for the connection or attachment, or both, of components.

Landscape Mode
(1) In facsimile, the mode for scanning lines across the longer dimension of a rectangular original. (2) In computer graphics, the orientation of a page in which the longer dimension is horizontal.

Lap Joint
A joint between two overlapping members.

R. Hanifan, *Concise Dictionary of Engineering: A Guide to the Language of Engineering,*   151
DOI 10.1007/978-3-319-07839-7_12, © Springer International Publishing Switzerland 2014

Laser Welding
A welding process that uses the concentrated energy of a focused laser beam to obtain fusion.

Launcher Dispersion
The departure (usually, but not necessarily, random) from the desired flight path that a guided missile takes during the launching phase.

Launcher, Rail-Type
A structure supporting a set of rails that in turn support the missile-booster combination. The rails provide orientation and control during the early portion of the launching phase.

Launcher, Zero Length
A launcher that supports the missile in the desired attitude prior to ignition, but that exercises little control on the direction of the missile's travel after ignition.

Launching Rail
A rail that gives initial support and guidance to a rocket launched in a nonvertical position.

Launch System
A composite of equipment, skills, and techniques capable of launching and boosting the space vehicle into orbit. The launch system includes the space vehicle, the launch vehicle, and related facilities, equipment, material, software, procedures, services, and personnel required for their operation.

Lay
Lay is the direction of the predominant surface pattern, ordinarily determined by the production method used.

Layout Characteristics
A description of the elements (i.e., page, block) of a document and the relationship between the elements.

Layout Drawing
A layout drawing depicts design development requirements. It is similar to a detail, assembly, or installation drawing, except that it presents pictorial, notational, or dimensional data to the extent necessary to convey the design solution used in preparing other engineering drawings.

Layup
A laminate that has been assembled but not cured, or a description of the component materials and geometry of a laminate.

Lead Projection
The distance that a component lead protrudes through the printed board on the side opposite from which the component is installed.

Lead Thread
That portion of the incomplete thread that is fully formed at root but not fully formed at crest that occurs at the entering end of either external or internal threads.

Lead Wire
Refers to an insulated conductor forming an integral part of components, such as motors, transformers, hall generators, etc., and used for chassis wiring.

Leakage Current
All currents, including capacitively coupled currents that conduct between exposed conductive surfaces of a unit and ground or other exposed surfaces of the unit.

Least Material Condition (LMC)
The condition in which a feature of size contains the least amount of material within the stated limits of size. Example: maximum hole, minimum shaft diameter.

Left-Handed Thread
A screw thread that is screwed in moving counterclockwise. All left-handed threads are designated "LH."

Leg
A connection from a specific node to an addressable entity, which may be another network node, such as a switching system, but will ordinarily be a termination to a user of the network.

Length, Free
The overall length of a spring to which no external force has been applied.

Length of Thread Engagement
The axial distance over which two mating threads, each having full form at both crest and root, are designed to engage.

Length, Solid
The overall length of a compression spring when all coils are fully compressed.

Lens, Achromatic
A lens consisting of two or more elements, usually made of crown and flint glass, which has been corrected so that light of at least two selected wavelengths is focused at a single axial point.

Lens, Cartesian
A lens, one surface of which is a Cartesian oval. It produces an aplanatic condition.

Lens, Collective
A lens of positive power used in an optical system to refract the chief rays of image-forming bundles of rays so that these bundles will pass through subsequent optical elements of the system. If the entire bundles do not pass through an optical element, a loss of light ensues, known as *vignetting*. Sometimes the term *collective lens* is used incorrectly to denote any lens of positive power.

Lens, Converging
A lens that adds convergence to an incident bundle of rays. One surface of a converging lens may be con-vexedly spherical and the other plane (plano-convex), both may be convex (double-convex, biconvex), or one surface may be convex and the other concave (converging meniscus).

Lens, Diverging
A lens that causes parallel light rays to spread out. One surface of a diverging lens may be concavely spherical and the other plane (plano-concave), both may be concave (double concave), or one surface may be concave and the other convex (concave-convex, divergent-meniscus). The diverging lens is always thicker at the edge than at the center.

Lens, Plano
A lens having no curved surface, or whose two curved surfaces neutralize each other, so that it possesses no refracting power.

Lesson Learned
A proven experience of value in the conduct of future programs. It is normally a conclusion drawn from evaluation of feedback information or from analysis of the performance resulting from technical and management functional activities. A lesson learned is usually recorded and eventually incorporated, where applicable, in regulations, technical manuals, specifications, standards, or handbooks.

Levels of Protection
A means of specifying the minimum preservation and packing that a given item requires to assure that it is not degraded during shipment and storage.

Life Cycle
A generic term covering all phases of acquisition, operation, and logistics support of an item, beginning with concept definition and continuing through disposal of the item.

Life Cycle Cost
The total cost to the Government of acquisition and ownership of that system over its life cycle. It includes the cost of development, acquisition, support, and (where applicable) disposal.

Light Emitting Diode (LED)
A pn-junction semiconductor device that emits incoherent optical radiation when biased in the forward direction.

Lightning Direct Effects
Any physical damage to the system structure and electrical/electronic equipment due to the direct attachment of the lightning channel. These effects include tearing, bending, burning, vaporization, or blasting of hardware.

Lightning Indirect Effects
Electrical transients induced by lightning in electrical circuits due to coupling of electromagnetic fields.

Limited Coordination Military Specification
A limited coordination military specification covers items of interest to a single military department or activity and is prepared to meet the acquisition needs of that department or activity.

Limited-Coordination Specification
A specification that covers items required by a single activity, Military Department, or Defense Agency, and is coordinated only within that activity, Military Department, or Defense Agency.

Limited Design Disclosure Model
A Computer Aided Design (CAD) 3-Dimensional model sufficiently defined to provide a visual understanding of the item, but which does not contain full design disclosure. Generally key interface characteristics and features such as weight and center of gravity will be sufficiently defined for the intended purpose. Sometimes referred to as a shrink-wrap, visualization, or cosmetic model

Limited Rights

Rights to use, duplicate, or disclose technical data, in whole or in part, by or for the Government, with the express limitation that such technical data shall not, without the written permission of the party asserting limited rights, be (a) released or disclosed outside the Government, (b) used by the Government for manufacture, or in the case of computer software documentation, for preparing the same or similar computer software, or (c) used by a party other than the Government, except that the Government may release or disclose technical data to persons outside the Government or permit the use of technical data by such persons.

Limiter

A device in which the voltage or some other characteristic of the output signal is automatically prevented from exceeding a specified value.

Limiter Circuit

A circuit of nonlinear elements that restricts the electrical excursion of a variable in accordance with specified criteria.

Limiting Quality

Limiting quality is the maximum defectiveness in product quality (or the worst product quality) that the consumer is willing to accept at a specified probability of occurrence.

Limits of Size

The specified maximum and minimum sizes.

Line Balance

The degree of electrical similarity of the two conductors of a transmission line.

Line Driver

A digital-signal amplifying device used to enhance transmission reliability over an extended distance.

Line Progression

The direction of progression of successive lines of pels in an image.

Line Replaceable Unit (LRU)

(1) A unit designated to be removed, upon failure, from a larger entity (equipment, system) in the operational environment. (2) A component, assembly, or subassembly that is normally removed and replaced as a single unit to correct a deficiency or malfunction. LRUs may be composed of shop replaceable units (SRUs), which are generally removed and replaced in a maintenance shop.

Liquid Crystal Display (LCD)

A segmented, solid state, passive display device consisting of a liquid crystal material, composed of specialized molecules, sandwiched between two conductive plates, at least one of which is transparent. Transmission of light through the medium containing the crystals is affected by the orientation of the crystals. When a current is applied, the orientation of the crystals, and therefore the transmission characteristics of the medium, are altered, resulting in contrast between particular segments/pixels and their background.

Linearity

The condition wherein the change in the value of one quantity is directly proportional to the change in the value of another quantity.

Load
(1) The power consumed by a device or circuit in performing its function. (2) A power-consuming device connected to a circuit. (3) To enter data or programs into storage or working registers. (4) To insert data values into a database that previously contained no occurrences of data. (5) To place a magnetic tape reel on a tape drive, or to place cards into the card hopper of a card punch or reader.

Lock-On
The instant at which radar is enabled to automatically track its target.

Logic Diagram
A drawing that depicts the multistate or two-state device implementation of logic functions with logic symbols and supplementary notations, showing details of signal flow and control, but not necessarily the point-to-point wiring.

Logic Element
An element whose input and output signals represent logic variables, and whose output or outputs are defined functions of the inputs and of the variables (including time dependencies) internal to the element.

Logic Function
A combinational, delay, or sequential logic element defined in terms of the relationships that hold between input and output logic variables.

Logic Symbol
The graphic representation in diagrammatic form of a log function.

Logic Variable
A variable that may assume one of two discrete states, the 1-state or the 0-state.

Logistic Support Analysis (LSA)
The selective application of scientific and engineering efforts undertaken during the acquisition process, as part of the system engineering and design processes, to assist in complying with supportability and other integrated logistics support objectives.

Logistic Support Analysis Record (LSAR)
That portion of logistic support analysis documentation consisting of detailed data pertaining to the identification of logistic support resource requirements of a system/equipment.

Look-Ahead Buffer
A buffer that contains the data currently being analyzed by the compression algorithm.

Look-Thru
A visual display that continuously monitors the target signal in time and frequency. It derives its primary importance from the desirability for rapid acquisition, setting-on, and frequency of a communication signal while jamming.

Loose or Unpacked Item
An identifiable item that is unencumbered by a tie, wrap, or container.

Loran
Long-range electronic navigation system that uses the time divergence of pulse-type transmissions from two or more fixed stations. Also called *long-range navigation*.

Lot or Batch

(1) The term *lot* or *batch* shall mean "inspection lot" or "inspection batch," i.e., a collection of units of product from which a sample is to be drawn and inspected. It may differ from a collection of units designated as a lot or batch for other purposes. (2) For fibers and resins, a quantity of material formed during the same process and having identical characteristics throughout. For prepregs, laminae, and laminates, material made from one batch of fiber and one batch of resin. (3) An identifying number assigned by the enterprise to a designated group of items referred to as a lot. A lot may be further subdivided as batch.

Lot or Batch Number

An identifying number assigned by the enterprise to a designated group of items referred to as a lot. A lot may be further subdivided as batch. (a) TEI LOT Is a lot number that *is not unique* within the Enterprise Identifier but is unique within the Original Part Number TEI (PNO) (b) TEI LTN Is a lot number that *is unique* within the Enterprise Identifier. Referred to as Enterprise Lot Number.

Lower Frequency Ground

A dedicated, single-point network intended to serve as a reference for voltages and currents, whether signal, control, or power, from DC to 30 kHz and, in some cases, to 300 kHz. Pulse and digital signals with rise and fall times greater than 1 microsecond are considered to be lower frequency signals.

Low Order Mode

A propagation path that makes a relatively small angle with respect to the fiber axis.

M

Machine Language
A language, using sequences of 0s and 1s to convey information and instructions to a computer, and requiring no translation prior to interpretation by the computer.

Machine-Readable Information (MRI) Marking
A pattern of bars, squares, dots, or other specific shapes containing information interpretable through the use of equipment specifically designed for that purpose. The patterns may be applied for interpretation by digital imaging, infrared, ultra-violet, or other interpretable reading capabilities.

Mach Number
The ratio of the velocity of a body to that of sound in the surrounding medium.

Macro
(1) The capability to allow the user to assign a single name or function key to a defined series of commands for use with subsequent command entry. Sometimes called a *smart key*. Examples of use are storage of addresses or signature blocks that are frequently used. (2) In relation to composites, denotes the gross properties of a composite as a structural element but does not consider the individual properties or identify of the constituents.

Magnetic Card
A card with magnetizable layer on which data can be stored.

Magnetic Circuit
(1) The complete closed path taken by magnetic flux. (2) A region of ferromagnetic material, such as the core of a transformer or solenoid, that contains essentially all of the magnetic flux.

Magnetic Disk
A flat, circular plate with a magnetizable surface layer, on one or both sides, on which data can be stored.

Magnetic Drum
A right circular cylinder with a magnetizable layer on which data can be stored.

Magnetic Tape
(1) A tape with a magnetizable layer on which data can be stored. (2) A tape or ribbon of any material impregnated or coated with magnetic or other material on which information may be placed in the form of magnetically polarized spots.

Maintainability
(1) The measure of the ability of an item to be retained in or restored to specified conditions when maintenance is performed by personnel having specified skill levels, using prescribed procedures and resources, at each prescribed level of maintenance and repair. (2) A measure of the ease and rapidity with which a system or equipment can be restored to operational status following a failure or retained in a specified condition. It is characteristic of equipment design and installation, personnel availability in the required skill levels, adequacy of maintenance procedures and test equipment, and the physical environment under which

maintenance is performed. One expression of maintainability is the probability that an item will be retained in or restored to a specified condition within a given period of time when the maintenance is performed in accordance with prescribed procedures and resources.

Maintainability, Design for
Design considerations directed toward achieving those combined characteristics of equipment and facilities that will enable the accomplishment of necessary maintenance, quickly, safely, accurately, and effectively with minimum requirements for personnel, skills, special tools, and cost.

Maintenance
Activity intended to keep equipment (hardware) or programs (software) in satisfactory working condition, including tests, measurements, replacements, adjustments, repairs, program copying, and program improvement. Maintenance is either preventive or corrective.

Maintenance Concept
A description of the general scheme for maintenance and support of an item in the operational environment.

Maintenance Element
A discrete portion of a maintenance task that can be described or measured.

Maintenance Engineering Analysis
A process performed during the development stage to derive the required maintenance resources such as personnel, technical data, support equipment, repair parts, and facilities.

Maintenance Level
(1) The level at which maintenance is to be accomplished; that is, organizational, intermediate, and depot. (2) One of several organizational entities to which materiel maintenance functions may be assigned. The maintenance levels are unit, intermediate, and depot.

Maintenance (Repair) Parts Interchangeability
Maintenance (repair) parts interchangeability involves the installation and operation of a maintenance part in an item in life of a like item without the use of additional tools or modifications to the existing item or mounting facilities and with no appreciable effect on performance or ratings, either electrical or mechanical.

Major Defect
A defect, other than critical, that is likely to result in failure or to reduce materially the usability of the unit of product for its intended purpose.

Major Defective
A unit of product that contains one or more major defects; may also contain minor defects, but contains no critical defect.

Major Diameter
On a straight thread, the major diameter is that of the major cylinder. On a taper thread, the major diameter at a given position on the thread axis is that of the major cone at that position.

Major Lobe
The major lobe is the portion of the antenna field pattern containing the direction of maximum radiation.

Major Nonconformance
A nonconformance, other than critical, that is likely to result in failure or to materially reduce the usability of the supplies or services for their intended purpose.

Malfunction
(1) A physical condition that causes a device, a component, or an element to fail to perform in a required manner; for example, a short or open circuit or an intermittent connection. (2) A degradation in performance due to detuning, maladjustment, misalignment, or failure of parts. (3) Immediate cause of failure (e.g., maladjustment, misalignment, defect, etc.).

Malware
Software that is harmful or evil in intent.

Managing Activity
The organizational element of DoD assigned acquisition management responsibility for the system, or prime or associate contractors or subcontractors who wish to impose system safety tasks on their suppliers.

Man-Day
 - see MAN UNITS.

Man-Hour
 - see MAN UNITS.

Man-Month
 - see MAN UNITS.

Manpower and Personnel Integration (Manprint)
A program whose purpose is to impose human factors, manpower, personnel, training, system safety, and health hazard considerations across the entire materiel acquisition process.

Manufactured Head
The head formed at the time a rivet is manufactured.

Manufacturer (MFR)
An individual, company, corporation, firm or Government activity that controls the production of an item, or produces an item from crude or fabricated materials, or assembles materials or components, with or without modification, into more complex items.

Manufacturer's Identification
The actual manufacturer's name and enterprise identifier that identifies the place of manufacture (see also Enterprise Identifier).

Manufacturing Baseline
The manufacturing baseline is a description, normally in the form of a flow chart, of the sequence of manufacturing operation necessary to produce a specific item, part, or material. The manufacturing baseline includes all associated documentation that is identified or referenced, such as: that pertaining to the procurement and receiving inspection, storage, and inventory control of parts and materials used; the manufacturing processes; the manufacturing facilities, tooling, and test equipment; the in-process manufacturing controls; the operator training and certification; and the inspection and other quality assurance provisions imposed. Each document is identified by title, number, date of issue, applicable revision, and date of revision.

Man Units
A concept used to estimate or measure human energy to be expended or which has been expended on a particular project. The concept is ultimately based on the length of a working day, 6 or 8 h (productive time or calendar time). Thus, if a man-day is 6 h, 5 days or 30 man-hours is a man-week; 48 man-weeks or 1,440 man-hours is a man-year; 4 man-weeks or 120 man-hours is a man-month. A similar set of correspondences can be constructed based on 8 h (or any other number) per day.

Man-Year
- see MAN UNITS.

Maraging Steel
A steel containing nickel as the principal alloying addition (with lesser amounts of Co and Mo) that can be heat treated to a tensile strength of 200,000 psi or more by aging between 455 and 510°C.

Marginal Testing
Testing the current results on an indicator that has tolerance bands for evaluating the signal or characteristics being tested. (For example, a green band might indicate an acceptable tolerance range representing marginal operation, and a red band a tolerance that is unsatisfactory for operation of the item.)

Margins
The difference between the subsystem/equipment design level and the subsystem/equipment stress level.

Martensitic Steel
A stainless steel containing more than 12% chromium (and up to 2.5% Ni) that can be hardened by quenching and tempering.

Mask
- see RESIST.

Master Drawing
A document that shows the dimensional limits or grid location applicable to any or all parts of a printed board (rigid or flexible), including the arrangement of conductive and nonconductive patterns or elements, size, type, and location of holes; and any other information necessary to describe the product to be fabricated.

Master Film
Any film, but generally a negative, that is carefully inspected or edited and then used only in making prints. A master film is never used for day-by-day reference.

Master Positive
A picture or sound duplicate print, usually made for the purpose of producing a picture or sound duplicate negative for release printing.

Match Draft
Additional draft allowance permitted on matching surfaces at parting lines when the normal draft allowance would result in an offset of the surfaces at the parting lines.

Matched Metal Molding
Method of molding reinforced plastics between two close-fitting metal molds mounted in a hydraulic press.

Matched Parts
Those parts, such as special application parts, that are machine or electrically matched, or otherwise mated, and for which replacement as a matched set or pair is essential.

Matched Set Drawing
A matched set drawing delineates items that are matched and for which replacement as a matched set is essential.

Material Lot
A lot for material refers to material produced as a single batch or in a single, continuous operation or production cycle and offered for acceptance at any one time.

Material Review Board (MRB)
(1) The formal Contractor-Government Board established for the purpose of reviewing, evaluating, and disposing of specific nonconforming supplies or services, and for assuring the initiation and accomplishment of corrective action to preclude recurrence. (2) A board consisting of representatives of contractor departments necessary to review, evaluate, and determine or recommend disposition of nonconforming material referred to it.

Material Specification
A type of program-unique specification that describes such raw or processed materials as metals, plastics, chemicals, synthetics, fabrics, and any other material that has not been fabricated into a finished part or item.

Matrix
The essentially homogeneous material in which the fiber system of a composite is embedded.

Maximum Material Condition (MMC)
The condition in which a feature of size contains the maximum amount of material within the stated limits of size. Example, minimum hole diameter, maximum shaft diameter.

Maximum Operating Pressure
The maximum operating pressure is the highest pressure that can exist in a system or subsystem under normal (noncasualty) operating conditions. This pressure is determined by such influences as pressure regulating valve set pressure, maximum pressure at the system source such as compressed gas bank pressure or sea pressure, and pump or compressor shut-off pressure for closed systems. For constant pressure systems, such as regulated compressed gas systems, the term Nominal Operating Pressure has been used to designate the steady state operating condition. Where this type of pressure rating is applied, for the purpose of the requirements stated herein, it may be substituted as the maximum operating pressure.

May
"Should" and "may" are used when it is necessary to express nonmandatory provisions.

Mean Time Between Failures (MTBF)
(1) For a particular interval, the total functioning life of a population of an item divided by the total number of failures within the population during the measurement interval. The definition holds for time, cycles, miles, events, and other measure of life units. (2) A basic measure of reliability for repair items; the mean number of life units during which all parts of the item perform within their specified limits, during a particular measurement interval under stated conditions.

Measling
An internal condition occurring in laminated base material in which the glass fibers are separated from the resin at the weave intersection. This condition manifests itself in the form of discrete white spots or "crosses" below the surface of the base material, and it is usually related to thermally induced stress.

Measurement Sensitive Specification
A measurement sensitive document is one in which application of the requirements depends substantively on some measured quantity (for example, the document contains requirement for dimensions that are critical to the interfacing of the item).

Mechanical Interchangeability
The modified item must be capable of being physically installed and operated in the position previously occupied by the basic or previous item without requiring any major modifications. Switches, connectors, etc., shall be in the same location, within allowable tolerances. The center of gravity of the new item shall be the same as in the old item within allowable tolerances.

Mechanical Splice
An optical-fiber splice accomplished by fixtures or materials rather than by thermal fusion. *Note*: index-matching material may be applied between two fiber ends.

Median Line, Derived
An imperfect (abstract) line formed by the center points of all cross sections of the feature. These cross sections are normal (perpendicular) to the axis of the unrelated actual mating envelope.

Median Plane, Derived
An imperfect (abstract)plane formed by the center points of all line segments bounded by the feature. These line segments are normal (perpendicular) to the center plane of the unrelated actual mating envelope.

Megahertz (MHz)
A unit of frequency denoting one million hertz.

Message Synchronization
The ability to maintain message sequence continuity during message transfers.

Metacompiler
A compiler system designed specifically to implement (compile) language compilers.

Metadata
Metadata is generally defined to be data about data. Metadata is defined to be data about a design and its defining documents/models. Metadata is used by the Procuring Activity to store, manage, and provide access to TDP elements.

Metalanguage
A metalanguage is a formal mechanism used to describe, specifically, other languages.

Metal-Arc Welding
Any of the arc-welding processes in which fusion is obtained by heating with an arc between a consumable metal electrode and the work.

Metal Transfer
The transfer of filler metal from an electrode or welding rod to the work.

Meta-Programming
The process of expressing problems in an extended meta-language.

Meter
A unit of metric measurement. 1,000 millimeters equals one meter. 100 centimeters equals one meter. One meter equals 39.37 inches.

Metric
(1) A measure of the extent or degree to which the software possesses and exhibits a certain characteristic, quality, property, or attribute. (2) A meaningful measure of the extent or degree to which an entity possesses or exhibits a particular characteristic.

Metrication
Metrication is the process of changing to the metric system, including the act of developing metric standardization documents or converting current standardization documents to metric units of measurement.

Metric Design
Product design using metric dimensions, selected as appropriate, without considering conceptual or physical conversion from inch-pound units.

Metric, Metric System, Metric Units
The international System of Units (commonly abbreviated as SI), as established by the General Conference of Weights and Measures in 1960 and as interpreted or modified for the United States by the Secretary of Commerce.

Metrology
The science of measurement for determination of conformance to technical requirements including the development of standards and systems for absolute and relative measurements.

Micro
In relation to composites, denotes the properties of the constituents, i.e., matrix and reinforcement and interface only, as well as their effects on the composite properties.

Microcode
A sequence of microinstructions that is fixed in storage that is not program-addressable, and that performs specific processing functions.

Microcomputer
A computer system whose processing unit is a microprocessor. A basic microcomputer includes a microprocessor, storage, and an input/output facility, which may or may not be on one chip.

Microelectronic Device
An item of inseparable parts and hybrid circuits usually produced by integrated circuit techniques.

Microelectronics
Microelectronics is that area of electronic technology associated with or applied to the realization of electronic systems from extremely small electronic parts or elements.

Micro Image
An image located within a developed frame of microfilm too small to be read without magnification. The image usually depicts textual or drawing sheet information.

Microprocessor
A single LSI circuit that performs the functions of a CPU. Some characteristics of a processor include small size, inclusion in a single integrated circuit or a set of integrated circuits, and low cost.

Micro Program
(1) A program implemented in microcode. (2) A sequence of instructions, hardwired in a computer and operating on individual bits of digital words, that the computer uses to interpret machine language instructions.

Militarized
Those items that are specified and manufactured to military specifications and shall withstand all environmental conditions that may be encountered during wartime service.

Military Levels of Packing
The packing levels are level A, which provides maximum protection to meet the most severe worldwide shipment, handling, and storage conditions; and level B, which provides protection to meet moderate worldwide shipment, handling, and storage conditions.

Military Preservation
Preservation designed to protect an item during shipment, handling, indeterminate storage, and distribution to consignees worldwide

Military Specification
(1) A military specification covers systems, subsystems, items, materials, or products that are intrinsically military in character or are used in, or in support of, weapons systems and involve an essential system function or interface. (2) A document intended primarily for use in procurement, which clearly and accurately describes the essential technical requirements for items, materials, or services, including the procedures by which it will be determined that the requirements have been met. Specifications for items and materials may also contain preservation-packaging, packing, and marking requirements.

Military Standard
A document that establishes engineering and technical requirements for items, equipments, processes, procedures, practices, and methods that have been adopted as standard.

Millimeter
A unit of metric measurement. 1,000 millimeters equal one meter. 10 millimeters equal one centimeter. 25.40 millimeters is one inch.

Millimicron
A unit of length in the metric system equal to 0.001 micron. It is also equivalent to 10 angstroms.

Minimal Acceptance
The least strict performance criteria that shall be met for equipment to be used in the application area for which the requirement is specified. Minimal acceptance is typically applied to commercial off the shelf (COTS) equipments for light duty (non-mission critical) applications.

Minor Defect
A minor defect is a defect that is not likely to reduce materially the usability of the unit of product for its intended purpose, or is a departure from established standards having little bearing on the effective use or operation of the unit.

Minor Diameter
On a straight thread, the minor diameter is that of the minor cylinder. On a taper thread, the minor diameter at a given position the thread axis is that of the minor cone at that position.

Minor Nonconformance
A nonconformance that is not likely to materially reduce the usability of the supplies or services for their intended purpose, or is a departure from established standards having little bearing on the effective use or operation of the supplies or services. *Note*: Multiple minor nonconformances, when considered collectively, may raise the category to a major or critical nonconformance.

Mishap
An unplanned event or series of events that results in death, injury, occupational illness, or damage to or loss of equipment or property.

Mismatch
The offset of features on a part caused by misalignment of opposing segments of a mold or die.

Missile Checkout
Performance of those procedures that enable a determination to be made of whether all parts of the missile are apparently capable of functioning properly.

Mission Critical
Equipment that contributes significantly to the safety, maneuverability, and continued mission capability of the platform.

Mission Critical Computer Resources (MCCR)
Computer resources acquired for use as integral parts of weapons; command and control; communications; intelligence; and other tactical or strategic systems aboard ships, aircraft, and shore facilities and their support systems. The term also includes all computer resources associated with specific program developmental test and evaluation, operational test and evaluation, and post-deployment software support including weapon system trainer devices, automatic test equipment, land-based test sites, and system integration and test environments.

Mission Profile
(1) A time-phased description of the events and environment an item experiences from initiation to completion of a specified mission, to include the criteria of mission success critical failures. (2) A plot of air distance versus altitude for a given configuration and liftoff gross weight with parameters of fuel and time. The relationship is based on a no-wind condition and a mission sequence of liftoff, thrust climb, and an approximate range cruise.

Mission Ready Status
Mission ready status is the status attained by a new item, modification, or installation (hardware, software, or both) when all required testing has been completed; required engineering, operation, and maintenance documentation have been received by the site; and test results show the item or modification is ready for mission support. Computer software attains this status after formal qualification review and when additionally the operational software configuration has been constructed and is ready for mission support.

Mission Reliability
The probability that, under stated conditions, a system or equipment will operate in the mode for which it was designed, i.e., with no malfunctions, for the duration of a mission, given that it was operating in this mode at the beginning of the mission.

Mission Stores
All stores excluding suspension and release equipment (carriage stores) are classified as mission stores. In general, these stores directly support a specific mission of an aircraft.

Mission Store Interface
The electrical interface on the mission store external structure where the aircraft or carriage store is electrically connected. This connection is usually on the mission store side of an umbilical cable.

Mission Time
The period of time in which an item must perform a specified mission.

Mixed Logic
When the 1-state of the variables is defined as the more positive or less positive of the two possible levels, depending upon the absence or presence of the polarity indicator symbol, mixed logic is used in the diagram.

Mnemonic
Assisting or intending to assist a human memory and understanding. Thus a mnemonic term is usually an abbreviation that is easy to remember: for example, mpy for multiply and acc for accumulator.

Mnemonic Code
A pseudo code in which information, usually instructions, is represented by symbols or characters that are readily identified with the information.

Mode
A way of operating a program to perform a certain subset of the functions that the entire program can perform, as selected by control data or operating conditions. Often, the mode of a program will be defined as program states, with transitions annotated to delineate events causing the passages between modes of operation.

Modem
Contraction of *modulator-demodulator*. A device that modulates and demodulates signals. *Note 1*: Modems are primarily used for converting digital signals into quasi-analog signals for transmission over analog communication channels and for reconverting the quasi-analog signals into digital signals. *Note 2*: Many additional functions may be added to a modem to provide for customer service and control features.

Modification
A major or minor change in the design of an item of materiel, performed to correct a deficiency, to facilitate production, or to improve operational effectiveness.

Modification Drawing
A modification drawing delineates changes to items after they have been delivered. When required for control purposes, a modification drawing shall require reidentification of the modified item.

Modification Letters
A modification letter is defined as a letter assigned in alphabetical sequence starting with the letter "A" to show a modification to a nomenclature equipment where at least one way interchangeability has been maintained. The modification letter follows the sequentially assigned number. For example, Receiver Radio ARC 150A/ARC is a modified version of the R-150/. ARC and is interchangeable with the original item.

Modular
Pertaining to the design concept in which interchangeable units are employed to create a functional end product.

Modularizing
The ability to remove and replace all components supporting a common function in a single operation.

Modulation
(1) A controlled variation of any property of a wave for the purpose of transferring information. (2) The process in which variation in amplitude, frequency, or phase of an oscillation is varied with time in accordance with the waveform of an intelligence signal, usually at the audio frequency rate at which the information is being transmitted.

Modulation Factor
In amplitude modulation, the ratio of the peak variation actually used to the maximum design variation in a given type of modulation.

Modulation Index
In angle modulation, the ratio of the frequency deviation of the modulated signal to the frequency of a sinusoidal modulating signal.

Modulation Rate
The reciprocal of the measure of the shortest nominal time interval between successive significant instants of the modulated signal.

Module
A combination of components, contained in one package or so arranged that together they are common to one mounting, that provide a complete function or function to the subsystems in which they operate.

Modulus of Elasticity
The ratio, within the elastic limit of a material, of stress to corresponding strain, i.e., the ratio of the change of a dimension to the force producing the change within the elastic limit of the material.

Moisture Content
The percentage of water in a finished material such as film, paper, wood, etc., expressed as percentage of original weight of the test sample.

Moisture Proof
(1) Not affected by moisture. (2) A barrier to moisture. Although materials that resist passage of moisture are often called *moisture proof*, their preferable designation is moisture barrier.

Mold
A form made of sand, metal, or other material into which material is poured or injected to produce a part.

Mold Line
A line generated by the theoretical intersection of projected surfaces.

Mold Release Agent
A lubricant applied to mold surfaces to facilitate release of the molded article.

Monocoque
A type of construction, as of a rocket body, in which all or most of the stress is carried by the skin. It may incorporate formers but not longitudinal stringers.

Mono Detail Drawing
A mono detail drawing delineates a single part. *Note*: A drawing detailing SHOWN and OPPOSITE parts using a single set of views is considered to be a tabulated monodetail drawing.

Motherboard
(1) A printed board assembly used for interconnecting arrays of plug-in electronic modules. (2) A Printed Circuit Board that is large as compared to the smaller Printed Circuit Boards that attach to it. The Motherboard holds smaller size boards called Expansion Cards, Daughter cards, or Mezzanine Cards.

Mounting Rail
A mounting rail is the surface of a rack on which panels mount.

Moving Target Indication (MTI)
A radar presentation that shows only targets that are in motion. Signals from stationary targets are subtracted out of the return signal by the output of a suitable memory circuit.

MS Sheet
A term that once meant "military standard sheet" but was changed many years ago to "military specification sheet." The term specification sheet includes MS sheet. It is a document that specifies requirements and verifications unique to a single style, type, class, grade, or model that falls within a family of products described under a general specification. New MS sheets are no longer being issued except for situations where it is necessary to expand a family of existing MS sheets. Specification sheets instead of MS sheets are now developed for all new general specifications, and when possible, replace existing MS sheets.

Mud Box
An unsheltered item of equipment that is sufficiently rugged to withstand adverse environments. It is expected to operate when exposed on the ground.

Multicast
A data burst destined to a subset of up to ten members of a network.

Multi Detail Drawing
A multi detail drawing delineates two or more uniquely identified parts in separate views or in separate sets of views on the drawing.

Multilayer Printed Board
The general term for completely processed printed circuit or printed wiring configurations consisting of alternate layers of conductive patterns and insulating materials bonded together, with conductive patterns, in more than two layers, and with the conductive patterns interconnected as required. The term includes both flexible and rigid multilayer boards.

Multipacting
The resonant flow of secondary electrons in a vacuum between two surfaces separated by a distance such that the electron transit time is an odd integral multiple of one half the period of the alternating voltage impressed on the surfaces.

Multiple Failures
Multiple failures are the simultaneous occurrences of two or more independent failures. When two or more failed parts are found during troubleshooting, and failures cannot be shown to be dependent, multiple failures are presumed to have occurred.

Multiplexer
A device that collects data from many individual sources and arranges the information for simultaneous transmission over a single network.

Multi-programmable Memory Device
A memory device containing software instructions or data that can be changed in the device once programmed.

Multi-start Thread
A screw thread with two or more threads. For this condition, pitch is equal to the thread lead divided by the number of thread starts.

Murphy's Law
If anything can go wrong, it will.

N

Nationally Recognized Standard
A specification or standard issued with the intent to establish common technical requirements. Such standards are developed by or for a Government activity or by a nongovernment organization (private sector association, organization, or technical society), which conduct professional standardization activities (plans, develops, establishes, or publicly coordinates standards, specification, handbooks, or related documents) and is not organized for profit.

National/NATO Stock Number (NSN)
A number assigned to each item of supply that is purchased, stocked, or distributed within the Federal Government or NATO.

National Stock Number (NSN)
A number assigned to each item of supply that is purchased, stocked, or distributed within the Federal Government or NATO.

NATO Commercial and Government Entity (NCAGE) Code
A five position alphanumeric code requiring an alpha in either the first or last position (e.g., AA123, 3AAAA, AAAA3, K2345 or 2345K), assigned to organizations located in North Atlantic Treaty Organization (NATO) member nations (excluding U.S.) and other foreign countries which manufacture and/or control the design of items supplied to a Government Military Activity or Civil Agency.

Natural Language
A type of dialog in which users compose control entries in a restricted subset of their natural language; for example, English, Japanese, Arabic.

Nautical Mile, Radar
The time interval, approximately 12.367 microseconds, required for radio frequency energy to travel one nautical mile and return: a total of two nautical miles.

Naval Tactical Data System (NTDS)
An automatic data processing system for combat ships within a fleet task force. The network is basically ship and airborne data links in communication and weapons systems.

Navigation, Electronic
Navigation by means of electronic equipment. The expression electronic navigation is more inclusive than navigation, since it includes navigation involving any electronic device or instrument.

Navigation, Inertial
Dead reckoning performed automatically by a device that gives a continuous indication of position by combining vectors for speed, direction, and other factors since leaving a starting point.

Near Field
The region of the field immediately surrounding an antenna where the inductive and capacitive fields predominate. In this region, the angular distribution of the field varies with distance from the antenna.

Negative Lens
A lens thinnest at the center, which causes light rays to diverge.

Negative Logic
When the 1-state of the variables is defined as the less positive of the two possible values, negative logic is used in the diagram.

Neper
A division of the logarithmic scale such that the number of nepers is equal to the natural logarithm of the scalar ratio of two currents or two voltages.

Network
(1) An organization of stations capable of intercommunications but not necessarily on the same channel. (2) Connection of two or more nodes; in a *computer network*, the specific nodes consist of computers, or processing or communications equipment.

Network, Attenuation
An arrangement of circuit elements, usually impedance elements inserted in circuitry to introduce a known loss or to reduce the impedance level without reflections.

Network, Balancing
Designed for use in a circuit in such a way that two branches of the circuit are made substantially conjugate; that is, such that an electromotive force inserted in one branch produces no current in the other branch.

Network, Crossover
A selective network that divides its audio output into two or more frequency bands for distribution to loudspeakers.

Network, Decoupling
A network of capacitors and chokes or resistors placed in leads that are common to two or more circuits to prevent unwanted and harmful interstage coupling.

Network, Dedicated
Point-to-point communications network that is not part of the Defense Communications System but has a preassigned purpose.

Network, Delta
An artificial network used in the case of a two-wire electric power supply enabling the symmetrical and asymmetrical voltages to be measured separately by means of three specified resistors, one connected between the two wires and one between each wire and earth.

Neutral
The AC power system conductor that is intentionally grounded on the supply side of the service disconnecting means. It is low-potential (white) side of a single-phase ACV circuit or the low-potential fourth wire of a three-phase Wye distribution system. The neutral provides a current return path for the AC power currents, whereas the grounding (or green) conductor should not, except during fault conditions.

Neutralization
A method of nullifying the voltage feedback from the output to the input circuits of an amplifier through the tube interelectrode impedances.

New Line
A control character that directs a printing or display device to position itself at the first print-ing or display position in the next line. That is, it causes the device to perform both the car-riage return and the line feed function. It is standard to assign new line the same code representation as line feed in ASCII and MSCII.

Nick
A nick is a small cut in the edge of a conductor.

Noble
Refers to the positive direction of electrode potential, thus resembling the noble metals (such as gold).

Node
(1) A terminal of any branch of a network or a terminal common to two or more branches of a network. (Also called *junction point, branch point,* or *vertex.*) (2) One of the set of discrete points in a flow graph. (3) An end point of any branch of a network or graph, or a junction common to two or more branches. (4) In a network, a point where one or more functional units interconnect transmission lines.

Nodule
A nodule is a rounded mass of irregular shape; a little lump.

Noise
Any electromagnetic disturbance with a useful frequency band, such as undesired electromag-netic waves in a transmission channel or device. When caused by natural electrical discharges in the atmosphere such as lightning, noise may be called *static.*

Nomenclature
The combination of an item name and a type designation or as designated in the contract or contract documents.

Nomenclature, Item Name
A name published in the Federal Cataloging Handbook or that name developed by the requestor in accordance with DOD-STD-100 (ASME Y14.100), the portion applicable to drawing titles. Item names used with type designator assignments will be consistent with the policies of the Federal Cataloging Program. *Examples of unacceptable item names include abbreviations, acronyms, descriptions of size frequencies, etc*

Nomenclature, Type Designator
A combination of letters and numbers arranged in a specific sequence to provide a short sig-nificant method of identification.

Nominal Cured Thickness
The nominal cured thickness is the thickness of a laminate or multilayer printed wiring board after the prepreg has been cured at the temperature and pressure specified for that particular class of resin flow.

Nominal Value
An assigned, specified, or intended value of any quantity with uncertainty in its realization.

Nonbasic Value
Attribute value that is only allowed in document interchange in the context of a given applica-tion profile if its use is declared in the document profile.

Noncompliant Printed Wiring Board
Printed wiring boards produced to acquisition documents that take exception to any of the requirements, inspections, and tests specified and in the applicable specification sheets that are not specifically detailed and approved in the QM plan.

Nonconformance
(1) The failure of a unit or product to conform to specified requirements. (2) The failure of a characteristic to conform to the requirements specified in the contract, drawings, specifications, or other approved product description.

Nonconformance, Critical
A nonconformance that judgment and experience indicate is likely to result in hazardous or unsafe conditions for individuals using, maintaining, or depending upon the supplies or services; or is likely to prevent performance of a vital agency mission.

Nonconformance, Major
A nonconformance, other than critical, that is likely to result in failure, or to materially reduce the usability of the supplies or services for their intended purpose.

Nonconformance, Minor
A nonconformance that is not likely to materially reduce the usability of the supplies or services for their intended purpose, or is a departure from established standards having little bearing on the effective use or operation of the supplies or services.

Nonconforming Material
Any item, part, supplies, or product containing one or more nonconformances.

Nonconforming Units
Units that do not conform to a specification or other inspection standard; sometimes called *discrepant* or *defective* units.

Nonconformities
Specific occurrences of a condition that does not conform to specifications or other inspection standards; sometimes called *discrepancies* or *defects*. An individual nonconforming unit can have the potential for more than one nonconformity (e.g., a door could have several dents and dings; a functional check of a carburetor could reveal any of a number of potential discrepancies).

Noncritical Failure
Any failure that degrades a performance or results in degraded operation requiring special operating techniques or alternative modes of operation that could be tolerated throughout a mission, but should be corrected immediately upon completion of mission.

Nondeliverable Software
Software that is not required to be delivered by the contract.

Nondevelopmental Software (NDS)
Deliverable software that is not developed under the contract but is provided by the contractor, the Government, or a third party. NDS may be referred to as reusable software, Government furnished software, or commercially available software, depending on its source.

Non-government Standard (or Document)
A national or international standardization document developed by a private sector association, organization, or technical society that plans, develops, establishes, or coordinates standards, specifications, handbooks, or related documents. The term does not include standards

of individual companies. Non-Government standards adopted by the DoD are listed in the ASSIST database.

Non-government Standardization Document
A standardization document developed by a private sector association, organization or technical society which plans, develops, establishes or coordinates standards, specification, handbooks or related documents.

Non-part Drawing
An engineering drawing that provides requirements, such as procedures or instructions, applicable to an item, when it is not convenient to include this information on the applicable part drawing.

Nonrecurring Costs
As applied to ECPs, these are onetime costs, which will be incurred if an engineering change is approved and which are independent of the quantity of items changed, such as cost of redesign, special tooling, or testing.

Nonrepairable Subassembly
A component, module, or subassembly is nonrepairable if the physical nature of the item is such that the item cannot be economically or feasibly repaired due to the excessive cost of material and labor required to affect such repair. This excessive cost is normally considered as 65%, or greater, of the acquisition cost of the item. Examples of nonrepairable assemblies are (a) an integrated circuit, (b) an encapsulated printed circuit card wherein the components and cards are sealed in a hard thermosetting plastic compound, and (c) a module wherein all printed circuit cards or components are sealed in a hard thermosetting plastic compound or the module is hermetically sealed.

Normalizing
A heat treating process in which nonair-hardening steels are air cooled from the austenitizing range to obtain a fine-grained metal structure.

Notice of Revision (NOR)
A document used to define revisions to drawings, associated lists, or other referenced documents that require revision after Engineering Change Proposal (ECP) approval.

Null
A point of minimum or zero indication; usually an indication that the quantity being measured has been balanced by an opposing quantity.

Number, Binary
A number expressed in binary notation.

O

Object Aperture
- see APERTURE, CLEAR.

Objective Quality Evidence
Any statement of fact, either quantitative or qualitative, pertaining to the quality of a product or service based on observations, measurements, or tests that can be verified. (Evidence will be expressed in terms of specific quality requirements or characteristics. These characteristics are identified in drawings, specifications, and other documents that describe the item, process, or procedure.)

Object Program
A computer program expressed in machine language, usually the result of translating a source program by an assembler or compiler.

Object Program Data
The resulting form of a source language program after processing by a compiler or assembler. They are also called *object modules*. The object program is in a format suitable for loading and execution. It may require additional processing by a link loader or link editor.

Oblique Projection
An oblique projection is one in which parallel projectors or lines of sight, make an angle other than 90 degrees with the plane of projection.

Observable Critical Item (OCI)
An OCI is any part or material specifically designed, selected, or qualified to meet specified observable requirements.

Observable Critical Process (OCP)
An OCP is any fabrication, manufacturing, assembly, installation, maintenance and repair, or other process or procedure that implements an observable design and satisfies observable system requirements.

Occlusion
An obstruction that prohibits one from viewing the image that should be displayed.

Occurrence
The first time a nonconformance is detected on a specific characteristic of a part or process. All nonconformances attributed to the same cause and identified before the date, item, unit, lot number, or other commitment for effective corrective action are also considered occurrences.

Octave Band
An octave band is a band of frequencies in which the ratio of the upper band-edge frequency to the lower band-edge frequency is equal to 2:1. The band-center frequencies are the preferred frequencies as specified in ASA-53. The width of an octave band in hertz is approximately 71% of its mid-band frequency.

Octet
A subdivision of bits numbered from 8 to 1, where bit 8 is the most significant bit and bit 1 is the least significant bit.

Off-Line
Descriptive of a system and of the peripheral equipment or devices in a system in which the operation of peripheral equipment is not under the control of the central processing unit.

Off-Line Testing
The testing of an item with the item removed from its normal operational environment.

Off-the-Shelf Item
An item that has been developed and produced to military or commercial standards and specifications, is readily available for delivery from an industrial source, and may be acquired without change to satisfy a military requirement.

One Hundred Percent Inspection
Inspection in which specified characteristics of each unit of product are examined or tested to determine conformance with requirements.

One-Line Diagram
  - see SINGLE-LINE DIAGRAM.

One-Time Programmable Memory Device
A memory device containing software instructions or data that cannot be changed in the device once programmed. Examples are programmable read only memory (PROM) and programmable array logic (PAL).

Online
A term used to describe a system with its peripheral equipment in which the operation of such equipment is under control of the central processing unit, and in which information reflecting current activity is introduced into the data processing system as soon as it occurs, thus, directly in-line with the main flow of transaction processing.

Opaque
(1) Pigmented material applied in suspension to portions of a photographic negative to prevent the passage of light during printing. Opaque portions will appear white in the print. (2) Not permitting transmission of light.

Open Cell (Sponge)
Open cell is a rubber-like product made by incorporating into the rubber compound an inflating agent, such as sodium bicarbonate, that gives off a gas that expands the rubber during vulcanization.

Open Harness
An assembly of wires and/or cables that does not include a protective outer covering.

Open Rack
An open rack consists only of the structural members necessary for supporting of equipment and is not intended to be enclosed. The vertical members provide mounting surfaces with holes for the mounting of panels.

Open System Architecture (OSA)
A system the implements sufficient open specifications, services, and supporting formats to enable properly engineered components to be utilized across a wide range of systems and

minimal changes, to interoperate with other components on local and remote systems, and to interact with users in a style that facilitates portability. An open system is characterized by (a) well defined, widely used, nonproprietary interfaces, (b) use of standards that are developed/adopted by industrially recognized standards bodies, (c) definition of all aspects of system interfaces to facilitate new or additional systems capabilities for a wide range of applications, and (d) explicit provision for expansion or upgrading through the incorporation of additional or higher performance elements with minimal impact on the system.

Operational Availability
The expected percentage of time that a weapon system or individual equipment will be ready to perform satisfactorily in an operating environment when called for at any random point in time.

Operational Capability
The extent to which a missile system can fulfill its assigned operational mission.

Operational Constraint
Limits (parameters) that define the operational characteristics and/or environment.

Operational Definition
Operational definition is a means of clearly communicating quality expectations and performance. It consists of (a) a criterion to be applied to an object or to a group, (b) a test of the object or group, and (c) a decision (yes or no) that the object or group did/did not meet the criterion.

Operational Readiness
(1) The ability of a military unit to respond to its operation plan upon receipt of an operations order. (2) The probability that, at any point in time, a system or equipment is either operating satisfactorily or ready to be placed in operation on demand when used under stated conditions, including stated allowable warning time. Thus, total calendar time is the basis for computation of operational readiness.

Operational Testing
A series of tests conducted by the designated user to determine operational effectiveness, suitability, and military desirability of materiel and the adequacy of the organization, doctrine, and tactics proposed for use.

Operational Value
A measure of maintainability that includes the combined effects of item design, quality, installation, environment, operation, maintenance, and repair.

Optical Attenuator
Absorbing or partially reflecting component placed in the optical beam to reduce the energy delivered to an optical detector.

Optical Carrier
The unmodulated component of light emitted by an optical source, suitable for modulation.

Optical Contact
An exact term in optics, signifying the approach of two transparent, index matched bodies so closely (a small fraction of a wavelength) that the discontinuity no longer produces a reflection. Typically, the surfaces of the two bodies must be polished to match each other's form before optical contacting can occur.

Optical Detector
A transducer that generates an output signal when irradiated with optical power.

Optical Fiber
Any filament of fiber made of dielectric a material that guides light, whether or not it is used to transmit signals.

Optical Fiber Cable
A fiber, multiple fibers, or fiber bundle in a structure fabricated to meet optical, mechanical, and environmental specifications.

Optimal Visual Zone
Refers to a 30° cone symmetrical about a line from the design eye position extending outward to the center of the instrument panel, the apex of the cone being at the design eye.

Organic
A material or compound composed of hydrocarbons or their derivatives, or those materials found naturally or derived from plant or animal origin.

Organizational Maintenance
That maintenance authorized for the responsibility of and performed by a using organization on its assigned equipment. In aeronautical systems, organizational maintenance normally consists of pre-flight, post-flight, and periodic inspection of aircraft; daily or minor inspection of other material; servicing; preventive maintenance; calibration of systems; and removal and replacement of components.

Original
The current design activity's full-size reproducible drawing or digital data file on the life of the drawing for historical record purposes.

Original Design Activity (ODA)
An activity (Government or contractor) having had responsibility originally for the design of an item and whose drawing number and CAGE Code is shown in the title block of drawings and associated documents.

Orthographic Projection
Orthographic projection is a system of drawing composed of images of an object formed by projectors from the object perpendicular to desired planes of projection.

Orthographic View
An orthographic view is the figure outlined upon the projection plane by means of the system of orthographic projection. Such a view shows the true shape of a surface parallel to the projection plane.

Outgassing
De-aeration or other gaseous emission from a printed board assembly when exposed to reduced pressure, heat, or both.

Output
(1) The current, voltage, power, or driving force delivered by a circuit or device. (2) The terminals or other places where current, voltage, power, or driving force may be delivered by a circuit or device.

Oven Soldering (OS)
A soldering process in which the heat required is obtained from an oven.

Overall Layout Drawings

System design drawings that include but are not limited to (a) the configuration and arrangement of major items of equipment for manned stations, such as a pilots or astronaut's station, or launch control officer's station, or shipboard command station; (b) the configuration and arrangement of items of equipment, such as modular rack or maintenance ground equipment, which may not be a part of a manned station for operation, but require human access for maintenance; (c) the arrangement of interior lighting for operating or maintaining the equipment; and (d) labels identifying general panel content, e.g., flight mission panel, countdown status panel, communications panel, or malfunction status panel.

Overhaul

To restore an item to a completely serviceable condition as prescribed by maintenance serviceability standards.

Overload

A load greater than that which a device is designed to handle. (It may cause overheating of power-handling components and distortion in a signal circuits.)

Overmodulation

More than 100% modulation. In amplitude modulation, overmodulation produces positive peaks of more than twice the carrier's original amplitude and brings about complete stoppage of the carrier on negative peaks, thus causing distortion.

Overshoot

The initial transient response to a unidirectional change in input which exceeds the steady-state response.

Over the Horizon Radar

Type of high-powered radar that uses scatter propagation to "see" over the horizon.

Oxy-Acetylene Welding

(1) A gas-welding process in which fusion is obtained by heating with a gas flame or flames obtained from the combustion of acetylene with oxygen, with or without the application of pressure or the use of filler metal. (2) A gas-welding process wherein coalescence is produced by heating with a gas flame or flames obtained from the combustion of acetylene with oxygen, with or without the application of pressure and with or without the use of filler metal.

Oxyhydrogen Welding (OHW)

A gas-welding process wherein coalescence is produced by heating with a gas flame or flames obtained from the combustion of hydrogen with oxygen, without the application of pressure and with or without the use of filler metal.

P

Package Content Drawing

A package content drawing is a drawing prepared to provide a package part identifying number and appropriate package nomenclature for stock identification of military material packaged for convenience of handling, storage, issue, or functional selectivity in logistic support operations. Package content drawings are prepared for packaging that constitutes a synthetic grouping or combination of items, which in themselves do not constitute a functioning, engineering, or production assembly.

Packaging

The processes and procedures used to protect materiel from deterioration, damage, or both. It includes cleaning, drying, preserving, packing, marking, and unitizing.

Packaging Density

Quantity of functions (components, interconnection devices, mechanical devices) per unit volume, usually expressed in qualitative terms, such as high, medium, or low.

Packet

A block of data often encoded with error correction and detection information so that each can be processed independently.

Packing

The assembly of items into unit packs and intermediate or exterior containers, with the necessary blocking, bracing, cushioning, weatherproofing, reinforcement, and marking.

Pad

- see LAND.

Pallet

A platform or skid on which lading is placed and secured. It is used to facilitate handling with mechanical handling equipment.

Palletized Unit Load

(a) A quantity of items, packed or unpacked, arranged on a pallet in a specified manner, and secured, strapped, or fastened on the pallet so that the whole palletized load is handled as a single unit. (b) Palletized or skidded load is not considered to be a shipping container. (c) A quantity of items, packed or unpacked, arranged on a pallet in a specified manner and is secured, strapped, or fastened on the pallet so that the whole palletized load is handled as a single unit.

Panel

A rectangular or square base material of predetermined size intended for or containing one or more printed boards and, when required, one or more test coupons.

Panel Layout Drawing

Equipment detail drawings that include, but are not necessarily limited to (a) a scale layout of the controls and displays on each panel or an item of equipment, such as an astronaut's, pilot's, or launch control officer's console; (b) a description of all symbols used; (c) identification of the color coding used for displays and controls; (d) the labeling used on each control or display; and (e) the identification of control type (e.g., alternate action or momentary) and a clear differentiation between controls and indicators.

R. Hanifan, *Concise Dictionary of Engineering: A Guide to the Language of Engineering*, DOI 10.1007/978-3-319-07839-7_16, © Springer International Publishing Switzerland 2014

Panning
An orientation for display framing in which a user conceives of the display frame as moving over a fixed array of data. The opposite of *scrolling*.

Panoramic Radar
Non-scanning radar that transmits signals over a side beam in the direction of interest.

Parallelism
Parallelism is the condition of a surface equidistant at all points from a datum plane or an axis equidistant along its length to a datum axis.

Parametric Device
Device whose operation depends essentially on the time-variation of a characteristic parameter usually understood to be a reactance.

Pareto Chart
A simple tool for problem-solving that involves ranking all potential problem areas or sources of variation according to their contribution to cost or to total variation. Typically, a few causes account for most of the cost (or variation), so problem-solving efforts are best prioritized to concentrate on the "vital few" causes, temporarily ignoring the "trivial many."

Parity
In binary-coded systems, the oddness or evenness of the number of ones in a finite binary stream.

Parity Check
A check that tests whether the number of ones or zeros in an array of binary digits is odd or even. *Note*: odd parity is standard for synchronous transmission, and even parity for asynchronous transmission.

Part
See also Basic Part. (1) One piece or two or more pieces joined together, which are not normally subject to disassembly without destruct or impairment of designed use. (2) An item made from a single piece of raw material or from multiple pieces joined together, which are not normally subject to disassembly without destruction or impairment of the designed use.

Part Design Failure
The failure of parts that can be traced directly to inadequate design.

Partial Reference Designation
A reference designation that consists of a basic reference designation and that may include, as prefixes, some but not all of the reference designations that apply to the assemblies or assemblies within which the item is located.

Partial Thread
 - see VANISH THREAD.

Partial View
Partial views may show only pertinent features not described by true projection in the principal views. They are used in lieu of complete views to simplify the drawing.

Particle Concentration
The number of individual particles per unit volume of air.

Particle Size
The apparent maximum linear dimension of a particle in the plane of observation as seen with a microscope, or the equivalent diameter of a particle detected by automatic instrumentation. The equivalent diameter is the diameter of a reference sphere having known properties and producing the same response in the sensing instrument as the particle being measured.

Parting
A selective attack of one or more components of a solid solution alloy e.g. dealuminization, dezincification (sometimes known as dealloying, or selective corrosion).

Parting Line
A line on a part corresponding to the separation between the mold or die segments.

Part Manufacturing Failure
A part failure that is the result of poor workmanship or inadequate manufacturing process control during part assembly, inadequate inspection, or improper testing.

Part Number (Pin) or Identifying Number
The identifier assigned by the responsible design activity or by the controlling nationally recognized standard that uniquely identifies (relative to that design activity) a specific item.

Part of
An item, required to complete the assigned function of an equipment, is "part of" that equipment. Items which are "part of" nomenclatured equipment are always listed in its complement data and issued automatically with the equipment of which it is a part.

Parts List (PL)
A tabulation of all parts and bulk materials (except those materials that support a process) used in the item.

Parts, Materials, and Processes Control Board (PMPCB)
The PMPCB is a formal contractor organization, formally established by contract direction to manage and control the selection, application, procurement, and documentation of parts, materials, and processes used in equipment, systems, or subsystems.

Passivation
Passivation is the treatment of the surface of corrosion-resistant steels with certain agents that will remove surface contaminants and produce a surface condition that is resistant to corrosive action.

Passive
A condition in which the behavior of a metal is more noble (less active) than its position in the EMF series would predict; a surface film protects the underlying metal from corrosion (opposite of *active*).

Passive Defense
Passive defense is the defense capability of a ship or shipboard equipment derived from its physical resistance to weapon damage effects and includes inherent structural strength, armoring, and certain design features incorporating mechanical protection, and special arrangements or redundancy of vital components.

Passive Graphics
The use of a display terminal in a noninteractive mode, usually involving the use of such items as plotters and microfilm viewers.

Passive Mode
A method of operation of a display device that does not allow any online interaction or alteration.

Passive Network
A network that does not include a power source.

Paste Flux
A flux formulated in the form of a paste to facilitate its application.

Patent Defect
An inherent or induced weakness that can be detected by inspection, functional test, or other defined means.

Patina
A green coating of corrosion products that grows on copper and some of its alloys after long exposure to the atmosphere.

Pattern Failures
The occurrence of two or more failures of the same part in identical or equivalent applications when the failures are caused by the same basic failure mechanism and the part failures occur at a rate that is inconsistent with the part's predicted failure rate.

Pattern Locating Tolerance Zone Framework (PLTZF)
The tolerance zone framework that controls the basic relationship between the features in a pattern with that framework constrained in translational and rotational degrees of freedom relative to the referenced datum features.

Peel, Orange
A term used to describe the uneven or dimpled appearance of a lens surface that has been improperly or insufficiently polished. A polished surface showing a granular appearance under magnification.

Peel Strength
The force per unit width required to peel the conductor or foil from the base material.

Peening
The mechanical working of metals by means of hammer blows.

Pel
The smallest graphic element that can be individually addressed within a picture. Synonymous with *pixel*.

Pel Density
The number of pels per unit distance in a raster image.

Pel Path
The direction of progression of successive pels along a line in an image.

Pel Spacing
The distance between any two successive pels along a scan line of an image. It is the inverse of such often-used terms as *pel density*, *transmission density*, and *resolution*.

Penetration
The minimum depth a groove or flange weld extends from its face into a joint, exclusive of reinforcement.

Penetration Aids
Components of the re-entry system utilized to penetrate and confuse enemy defenses, thus enhancing the mission probability of delivering a re-entry vehicle to its intended destination.

Penetration Ballistics
The part of terminal ballistics that treats the motion of a projectile, such as a guided missile warhead, as it forces it way into targets of solid or semisolid substances.

Penetration Velocity
The minimum velocity at which a particular projectile, such as a guided missile warhead, is expected to consistently perforate plates of given thickness and physical properties at a specific angle of obliquity.

Percent Defective
The percent defective of any given quantity of units of product is 100 times the number of defective units of product contained therein divided by the total number of units of product.

Percent Defective Allowable (PDA)
The percent defective allowable of a production lot of parts or materials is the maximum allowable percentage of parts or material specimens that fail to pass one or more tests before the entire production lot is considered to be unacceptable.

Percussion Welding (PEW)
A resistance welding process wherein coalescence is produced simultaneously over the entire abutting surfaces by the heat obtained from an arc produced by a rapid discharge of electrical energy with pressure percussively applied during or immediately following the electrical discharge.

Performance Specification
A specification that states requirements in terms of the required results with criteria for verifying compliance, but without stating the methods for achieving the required results. A performance specification defines the functional requirements for the item, the environment in which it must operate, and interface and interchangeability characteristics. Both defense specifications and program-unique specifications may be designated as a performance specification.

Performing Activity
The activity (Government, contractor, subcontractor, or vendor) that is responsible for performance of testability tasks or subtasks as specified in a contract or other formal document of agreement.

Periodic Maintenance
Maintenance performed on equipment on the basis of hours of operation or calendar time elapsed since last inspection.

Permanently Installed Part
A permanently installed part is defined as a detail part or assembly that is permanently installed as a part of the aircraft. Examples: rigid or whip antenna, bracket, cable assembly, fairlead, mounting, and plug.

Permanent Marking
Permanent marking is marking that is intended to remain integral with a component, component part, or piping for purposes of permanent identification.

Permanent Mask
A resist that is not removed after processing (e.g., plating resist used in the fully additive process).

Perpendicularity
Perpendicularity is the condition of a surface, media plane, or axis at a right angle to a datum plane or axis.

Personality Card
A printed circuit card that is inserted in a test adapter to make the test adapter compatible with I/O personality of a particular unit under test.

Perspective Projection
A perspective projection is one in which the projectors are not parallel and converge from points on the object to the point of sight located at a finite distance from the plane of projection.

pH
A measure of acidity or alkalinity of a solution. At 25°C, a neutral solution has a pH of 7; acid solutions have a pH <7, and alkaline solutions have a pH >7.

Phantastron
Electronic circuit of the multivibrator type, which is normally used in the monostable form. It is a stable trigger generator in this connection and is used in radar systems for gating functions and sweep delay functions.

Phased Array
Antenna term used to describe an array of dipoles on a radar antenna in which the signal feeding each dipole is varied in such a way that antenna beams can be formed in space and scanned very rapidly in azimuth and elevation.

Phase Lock Loop
A circuit that, normally, automatically controls an oscillator so that it remains in a fixed phase relationship with a reference signal. The phase lock loop is used in a variety of applications, such as tracking filters and frequency discriminators.

Phenolic Resin
A synthetic resin produced by the condensation of phenol with formaldehyde. Phenolic resins form the basis of a family of thermosetting molding materials, laminated sheet, and oven-drying varnishes. They are also used as impregnating agents and as components of paints, varnishes, lacquers, and adhesives.

Photodiode
A semiconductor device used in light-wave system to convert light energy to electrical energy.

Photographic Drawings
Pictorial illustrations for single parts and for exploded views prepared by photography.

Photographic Reduction Dimension
Dimensions, e.g., the distance between lines or between two specific points on the artwork master, used to indicate to the photographer the extent to which the artwork master is to be photographically reduced. (The value of the dimensions refers to the 1:1 scale and must be specified.)

Photon
The photon is the quantum (an indivisible energy unit) of the electromagnetic field. It has no electric charge, no rest mass, no magnetic field, and a long lifetime. In a vacuum, it travels at the speed of light.

Photo Resist
A photosensitive plastic coating material that hardens when exposed to ultraviolet light and resists etching solutions.

Photosensitive
Reactive to light energy. Photographic layers respond by darkening, by bleaching, by increasing their physical hardness, or more commonly by the formulation of a latent image. Photocells respond by generating an electric current. Certain chemicals respond by changing color, indicating changes in molecular structure.

Photostat®
A trademark of the Photostat Corp. for paper, chemicals, and equipment used in producing document copies on a photographic paper by means of a camera. The term is incorrectly applied to photocopies produced by materials and equipment of other origin.

Photostat Copy
Usually a black, positive reading photographic paper copy of any document made in a special machine. Also, white copies made from such black copies. See also Photostat.

Phugoid Oscillation
In a missile flight path, a long-period longitudinal oscillation consisting of shallow climbing and diving motions about a medium flight path and involving little or no change in angle of attack.

Physical Characteristics
Quantitative and qualitative expressions of material features, such as composition, dimensions, finishes, form, fit, and their respective tolerances.

Physical Characteristics Marking
The symbols, letters, numbers, color codes, and similar markings applied to indicate terminals, leads, and similar physical characteristics.

Physical Configuration Audit (PCA)
The formal examination of the "as-built" configuration of a configuration item against its technical documentation to establish or verify the configuration item's product baseline.

Physical Interchangeability
A condition in which any two or more parts or units made to the same specification can be mounted, connected, and used effectively in the same position in an assembly or system.

Pigtail
A short length of optical fiber permanently attached to an optical emitter, photodetector, or connector. It is used to couple power between the optoelectronic component and the transmission fiber.

Pilferable Items
Items that are vulnerable to theft because of their ready resale potential, such as cigarettes, alcoholic beverages, cameras, electronic equipment, and clothing and textiles.

Pin Photodiode
Positive intrinsic negative photodiode. In this light-sensitive semiconductor diode, the P-doped and N-doped regions are separated by an undoped "intrinsic" region. Pin photodiodes have the advantage of broad spectral response, wide dynamic range, high speed, and low noise, but no internal gain.

Pipe
Tube in standardized combinations of outside diameter and wall thickness, commonly designated by "Nominal Pipe Sizes" and "ANSI Schedule Numbers" (Note larger sizes usually greater than 1 inch are typically extruded while smaller sizes are typically drawn.)

Pit
(1) A depression in the conductive layer that does not penetrate entirely through it. (2) A term denoting small holes in a glass surface that can be seen as small, red particles by reflected light.

Pitch
(1) The pitch of a thread having uniform spacing is the distance, measured parallel to its axis, between corresponding points on adjacent thread forms in the same axial plane and on the same side of the axis. Pitch is equal to the lead divided by the number of thread starts. (2) The distance between adjacent active coils of a spring in the free position and measured at the material center.

Pitch Diameter
On a straight thread, the pitch diameter is the diameter of the pitch cylinder. On a taper thread, the pitch diameter at a given position of the thread axis is the diameter of the pitch cone at that position.

Pitch Fibers
Reinforcement fiber derived from petroleum or coal tar pitch.

Pitting
A very localized type of corrosion attack resulting in deep penetration at only a few sites (opposite of *general corrosion*).

Pixel
The smallest discrete scanning line sample of a facsimile system, which sample contains grayscale information. See also Pel.

Plan for Logistic Support
A major section of the materiel acquisition plan that deals with all aspects of materiel support planning.

Plane, Tangent
(1) A theoretically exact plane derived from the true geometric counterpart of the specified feature surface. (2) A plane that contacts the high points of the specified feature surface.

Plasma
A constricted electric arc-gas mixture.

Plasma-Arc Welding
An arc-welding process in which fusion is obtained by the heat of a constricted-arc plasma.

Plastic Deformation
A change in dimensions of an object under load that is not recovered when the load is removed; opposed to *elastic deformation*.

Plate
A rolled product that is rectangular in cross section and with thickness not less than 0.250 inch with sheared or sawed edges.

Plated-Through Hole
A hole in which electrical connection is made between internal or external conductive patterns, or both, by the deposition of metal on the wall of the hole.

Plating Bar
The temporary conductive path interconnecting areas of a printed board to be electroplated, usually located on the panel outside of the borders of such a board.

Plating Lot
A plating lot is defined as any number of printed wiring boards or composite panels that are placed in any one plating tank and are processed through that one particular plating cycle.

Plating Up
The process consisting of the electrochemical deposition of a conductive material upon the base material (surface holes, etc.) after the base material has been made conductive.

Plenum
Within a building, a space created by building components, designed for the movement of environmental air; e.g., a space above a suspended ceiling or below an access floor.

Plotting
The practice of mechanically converting X-Y positional information into a visual pattern, such as artwork.

Plug Weld
A circular weld made through a hole in one member of a lap or T-joint fusing that member to the other. The walls of the hole may or may not be parallel, and the hole may be partially or completely filled with weld metal.

Pogo Effect
The result of inharmonic surging of propellants in a guided missile fuel system, producing an extreme vertical force due to analomic pressures.

Point, Critical
A point in a subsystem considered most susceptible to interference due to sensitivity, inherent susceptibility, importance to mission objectives, or exposure to the electromagnetic environment. The critical point is electrical in nature and normally precedes the subsystem output stage.

Point, Monitor
Describes one or more points in a subsystem or system used to observe or measure responses of the subsystem or system. Monitor points for determining unacceptable response shall be at the system or subsystem output and need not be electrical in nature. Monitor points used in conjunction with critical points to determine that no inadvertent responses exist may be located at either internal system points or at the system or subsystem output. If monitor points are chosen at internal subsystem locations, particular caution must be exercised to ensure that the monitoring instrumentation does not influence the test results.

Polarity
The condition in an electrical circuitry by which the direction of the flow of current can be determined. Usually applied to batteries and other direct voltage sources.

Polarization
(1) A technique of eliminating symmetry within a plane so that parts can be engaged in only one way so as to minimize the possibility of electrical and mechanical damage or malfunction. (2) A shift in electrode potential from the open-circuit value as a result of current flow (also known as *overvoltage*).

Polyamide
A polymer in which structural units are linked by amide grouping. Many polyamides are fiber formers.

Polychlorinated Biphenyl (PCB)
An organic chemical, synthetically manufactured and used primarily in electrical equipment. It is harmful to human health and the environment.

Polyglycol
Materials with polyether linkages, such as polyethylene glycol or a material generally derived by the reaction of organic acids, amines, alcohols, phenols, or water with ethylene or propylene oxides, or their derivatives. This family of materials includes, but is not limited to, polyethylene glycol, polypropylene glycol, and a wide range of polyglycol surfactants. This family of materials does not include glycols (e.g., ethylene glycol); polyols (e.g., glycerin); or mono-, di-, or triglycol ethers.

Porosity
A condition of trapped pockets of air, gas, or vacuum within a solid material, usually expressed as a percentage of the total nonsolid volume to the total volume (solid plus nonsolid) of a unit quantity of material.

Port of Debarkation (POD)
An authorized point where shipments enter a country, either into the continental United States or into a foreign country.

Port of Embarkation (POE)
An authorized point where shipments leave a country, either from the continental United States or from a foreign country.

Positional Tolerance
A positional tolerance defines a zone within which the center, axis, or center plane of a feature of size is permitted to vary from true (theoretically exact) position. Basic dimensions establish the true position from specified datum features and between interrelated features. A positional tolerance is indicated by the position symbol, a tolerance, and appropriate datum references placed in a feature control frame. Previously known as *true position*.

Position Designation
User selection and entry of a position on a display, or of a displayed item.

Positive
An artwork, artwork master, or production master in which the intended conductive pattern is opaque to light, and the areas intended to be free from conductive material are transparent.

Positive-Acting Resist
A resist that is decomposed (softened) by light and that, after exposure and development, is removed from areas that were under the transparent parts of a production master.

Positive Lens
A lens thickest at the center, which causes light rays to converge.

Positive Logic
When the 1-state of the variables is defined as the more positive of the two possible values, positive logic is used in the diagram.

Pot Life
(1) The length of time that a resin system retains viscosity low enough to be used in processing. (2) The period of time during which a reacting thermosetting composition remains suitable for its intended processing after mixing with a reaction initiating agent.

Power Budget
A calculation of how much light energy must be provided by the transmitter to overcome various system losses and still satisfy the energy input requirements of the receiver.

Power Carrier
The average power supplied to the antenna transmission line by a transmitter during one radio frequency cycle under conditions of no modulation. This definition does not apply to pulse modulated emissions.

Power Density
The power density shall be defined as output power supply envelope volume, including cooling components/fins and EMI filtering where required.

Pre-award Survey (PAS)
An evaluation of a prospective contractor's capability to perform under the terms of a proposed contract.

Precipitation Hardening
Hardening caused by the precipitation of a constituent from a supersaturated solid solution.

Precipitation Heat Treatment
Artificial aging in which a constituent precipitates from a supersaturated solid solution.

Predictive Information
Information predicting future status, condition, or position of the aircraft, a system, or a subsystem.

Pre-emphasis
Systematic distortion of the speech spectrum to improve intelligibility of speech sound by attenuating the low-frequency components of vowels (relatively unimportant for intelligibility) and proportionately increasing the amplitude of high-frequency vowel components and consonants (highly important for intelligible speech transmission).

Preliminary Design Review (PDR)
This review shall be conducted for each configuration item or aggregated of configurations items to (a) evaluate the progress, technical adequacy, and risk resolution (on a technical, cost, and schedule basis) of the selected design approach, (b) determine its compatibility with performance and engineering specialty requirements of the configuration item development specification, (c) evaluate the degree of definition and assess the technical risk associated with the selected manufacturing methods/processes, and (d) establish the existence and compatibility of the physical and functional interfaces among the configuration item and other items of equipment, facilities, computer programs, and personnel.

Preparation Time
That element of active repair time required to obtain necessary test equipment and maintenance manuals and to set up the necessary equipment in preparation for fault location.

Preply
A composite material lamina in the raw material stage ready to be fabricated into a finished laminate. The lamina is usually combined with other raw laminae prior to fabrication. A preply includes all of the fiber system placed in position relative to all or part of the required matrix material that together will comprise the finished lamina. An organic matrix preply is called a *prepreg*.

Prepreg
(1) Sheet material (e.g., glass fabric) impregnated with a resin cured to an intermediate stage (B-stage resin). (2) Ready to mold or cure material in sheet form, which may be fiber, cloth, or mat impregnated with resin and stored for use. The resin is partially cured to a B-stage and supplied to the fabricator for layup and cure.

Preservation
The processes and procedures used to protect materiel against corrosion, deterioration, and physical damage during shipment, handling, and storage; application of protective measures, including cleaning, drying, preservative materials, barrier materials, cushioning, and containers when necessary.

Preventive Maintenance
The maintenance performed to retain an item in satisfactory operational condition by providing systematic inspection, detection, and prevention of incipient failures.

Preventive Maintenance Time
That portion of calendar time used in accomplishing preventive maintenance. It comprises time spent in performance measurement; care of mechanical wearout items; front panel adjustment, calibration, and alignment; cleaning, and scheduled replacement of items.

Primary Controls
The most important and frequently used devices designed to control equipment and systems.

Primary Display
The display that is most accessible to the user and usually the one most frequently used.

Primary Package
The unit container that is actually in contact with its contents.

Primary Repository
Repositories that receive, inspect, accept, and store the official record copy of data (camera masters), and disseminate copies of the master data to established customers and other appropriate requesters or users.

Prime Visual Signal Area (PVSA)
The PVSA is an area of the instrument panel that is enclosed by a circular arc whose radius is 12 inches and whose center is defined by the intersection of the top of the instrument panel and the crew member's vertical centerline plane. The area is the optimum location on the instrument panel for the pilot's flight instruments and the master caution and warning lights.

Printed Board
The general term for completely processed printed circuit and printed wiring configurations.

Printed Board Thickness
The overall printed wiring board thickness includes metallic depositions, fusing, and solder resist. The overall thickness is measured across the printed wiring board extremities (thickest

part), unless a critical area, such an edge card connector or card guide mounting location, is identified on the master drawing.

Printed Circuit

(1) A conductive pattern consisting of printed components (e.g., inductors, resistors, capacitors, etc.), printed wiring, or a combination of both, formed in a predetermined design and attached to a common base. (2) A circuit formed by depositing conducting material on the surface of an insulated sheet. Circuit components such as wiring, resistors, capacitors, inductors, and so forth are deposited or etched on the sheet by various processes. Also referred to as a *printed wiring board* (*PWB*).

Printed Wiring

A conductive pattern that provides point-to point connections, but no printed components, in a predetermined arrangement on a common base.

Printed Wiring Board Part Number

The term printed wiring board part number refers to a printed circuit or wiring board of a single specific part number and classification for a printed wiring board configuration. All samples of a printed board part number are to be electrically and functionally interchangeable with each other, have the same electrical and environmental test limits, and use the same basic raw materials, and fabrication processes.

Printed Wiring Board Test Specimen

The term printed wiring board test specimen is used to describe all of the following; production printed wiring boards, qualification test specimens, or test coupons.

Printed Wiring Board Tin Finishes

The use of alloys with tin content greater than 97% may exhibit tin whisker growth problems after manufacture. Tin whiskers may occur anytime from a day to years after manufacture, and can develop under typical operating conditions on products that use such materials. Tin whisker growth could adversely affect the operation of electronic equipment systems. Conformal coatings applied over top of a whisker-prone surface will not prevent the formation of tin whiskers. Alloys of 3% lead have shown to inhibit the growth of tin whiskers.

Probability of Acceptance

That percentage of inspection lots expected to be accepted when the lots are subjected to a specific sampling plan.

Probability of Fault Detection

By using authorized displays, manuals, checklists, test points, and test equipment, the probability that an existing fault that would render a system or equipment inoperable (or marginally effective) will be detected.

Probe

A short burst of information used to capture a channel or exchange control information.

Process

An operation, treatment, or procedure used during a step in the manufacture of a material, a part, or an assembly.

Process Average

The process average is the average percentage defective or average numbers of defects per hundred units (whichever is applicable) of product submitted by the supplier for original inspection. Original inspection is the first inspection of a particular quantity of product as distinguished from the inspection of product that has been resubmitted after prior rejection.

Processor
In a computer, a functional unit that interprets and executes instructions. *Note*: a processor consists of at least an instruction control unit and an arithmetic unit.

Process Specification
A type of program-unique specification that describes the procedures for fabricating or treating materials and items.

Procurement
The process of obtaining personnel, services, supplies, and equipment.

Procuring Activity
(1) A component of a Government agency having a significant acquisition function and designated as such by the head of the agency. Unless agency regulations specify otherwise, the term *procuring activity* shall be synonymous with *contracting activity*. (2) The procuring activity is the customer. This may include individual consumers, private industry, and Government organizations.

Product
Includes materials, parts, components, subassemblies, assemblies, and equipments. The term *product*, where used, shall also encompass a family of products. A family of products is defined as all products of the same classification, design, construction, material, type, etc. produced with the same production facilities, processes, and quality of material, under the characteristics defined and specified in the applicable engineering documentation.

Product Baseline (PBL)
The initially approved documentation describing all of the necessary functional and physical characteristics of the configuration item and the selected functional and physical characteristics designated for production acceptance testing and tests necessary for support of the configuration item. In addition to this documentation, the product baseline of a configuration item may consist of the actual equipment and software.

Product Configuration Documentation (PCD)
(1) The approved product baseline plus approved changes. (2) The combined performance/design documentation utilized for the production/procurement of the CI. The PCD incorporates the ACD describing a CI's functional, performance, interoperability, and interface requirements and the verifications required to confirm the achievement of those specified requirements. The PCD also includes such additional design documentation, ranging from form and fit information about the proven design to a complete design disclosure package, as is deemed necessary for the acquisition program.

Product Data
All engineering data, in processable form, necessary to define the geometry, the function, and the behavior of an item over its entire life span. The term includes logistic data elements for quality, reliability, maintainability, topology, relationship, tolerances, attributes, and data elements necessary to completely define the item for the purpose of design, analysis, manufacture, test, and inspection.

Product Definition Data
Denotes the totality of data elements required to completely define a product. Product definition data includes geometry, topology, relationships, tolerances, attributes, and features necessary to completely define a component part or an assembly of parts for the purpose of design, analysis, manufacture, test, and inspection.

Product Design Data
Product data that describes the physical configuration and performance characteristics of an item in sufficient detail to ensure that an item or component produced in accordance with the data will be essentially identical to the original item or component.

Product Drawings
Engineering drawings that provide the necessary design engineering, manufacturing, and quality support information necessary to permit a competent manufacturer to produce an interchangeable item that duplicates the physical and performance characteristics of the original design without additional design engineering or recourse to the original design activity.

Production Board
A printed board or discrete wiring board that has been manufactured in accordance with the applicable detail drawings, specifications, and procurement requirements.

Production and Deployment Phase
(1) The period from production approval until the last system is delivered and accepted. (2) The fourth phase in the materiel life cycle. During this phase, all hardware, software, and trained personnel required to deploy an operational system are acquired.

Production Lot
(1) A group of items manufactured under essentially the same conditions and processes. (2) A production lot of parts refers to a group of parts of a single part type; defined by a single design and part number; produced in a single production run by means of the same production processes, the same tools and machinery, same raw material, and the same manufacturing and quality controls; and to the same baseline document revisions and tested within the same period of time. All parts in the same lot have the same lot date code, batch number, or equivalent identification.

Production Master
A one-to-one scale pattern used to produce one or more printed boards (rigid or flexible) with the accuracy specified on the Master Drawing.

Production Model
An item, in its final mechanical and electrical form, of final production design made by production tools, jigs, fixtures, and methods.

Product Logistic Support Data
Product data that describes the equipment, tools, techniques, item characteristics, or analysis necessary to operate, maintain, or repair the item by its end user.

Product Manufacturing or Process Data
Product data that describes the steps, sequences, and conditions of manufacturing, processing, or assembly used by the manufacturer to produce an item or component or to perform a process.

Product Model Data
A three-dimensional geometric representation of a design that includes digital information required for full product definition. Additional attributes are provided to enable digital sharing, exchange and archiving of design, analysis, manufacturing, maintenance, repair, and reprocurement information.

Product Quality Review
An action by the Government to determine that the quality of supplies or services accepted by the Government do, in fact, comply with specified requirements.

Product Reliability Verification Test
A test to provide confidence that field reliability will be achieved.

Profile
A profile is the outline of an object in a given plane (two dimensional figure).

Profile (Unequally Disposed) (U)
In geometric dimensioning and tolerancing it is the symbol used to indicate a unilateral or unequally disposed profile tolerance. The symbol is placed in the feature control frame following the tolerance value.

Profile of Thread
The contour of a screw thread-ridge and groove delineated by a cutting plane passing through the thread axis; also called *form of thread*.

Program
(1) A separately compliable, structural (closed) set of instructions most precisely associated with early generations of computers. Synonymous with computer program. Contrast with software unit. (2) A program is a collection of operations or abstract entity designed to cause the computer equipment to execute an operation or operations. Computer programs include operating systems, assemblers, compilers, interpreters, data management systems, utility programs, as well as applications programs such as payroll, inventory control, operational flight, satellite navigation, automatic test, crew simulator, and engineering analysis programs. Computer programs may be either machine dependent or machine independent, and may be general purpose or be designed to satisfy the requirements of a specialized process or a particular user. (3) The lowest level of module that can be assembled or compiled and can be executed as a single entity. (4) An algorithm along with a particular collection of data objects to which the algorithm is applied. A program is taken as independent of the programming language in which it is expressed.

Programmable
Pertaining to a device that can accept instructions that alter its basic functions.

Programmable Logic Array
An array of gates whose interconnections can be programmed to perform a specific logical function.

Programmable Read-Only Memory (PROM)
A storage device that, after being written once, becomes a read-only memory.

Program-Unique Specification
A specification that describes a system, item, software program, process, or material developed and produced (including repetitive production and spares purchases) for use within a specific program, or as a part of a single system and for which there is judged to be little potential for use by other systems.

Projection Welding (RPW)
A resistance-welding process wherein coalescence is produced by the heat obtained from resistance to electric current through the work parts held together under pressure by electrodes. The resulting welds are localized at predetermined points by projections, embossments or intersections.

Project Management
The business and administrative planning, organizing, directing, coordinating, controlling, and approval actions designated to accomplish overall project objectives that are not associated with specific hardware elements and are not included in system engineering. Examples of

these activities are logistics management, cost/schedule/performance measurement, contract management, data management, vendor liaison, etc.

Prompt
An indicator provided by the computer that alerts the user that the computer is ready, data should be entered, etc.

Property Entity
A structure entity that allows numeric or test information to be related to other entities.

Protanope
An individual who exhibits protanopia, a severe type of color vision deficiency caused by the absence of the red retinal photoreceptors affecting hue discrimination in the orange-yellow-green section of the spectrum, and in which red appears dark.

Protected Cargo
Items that are required to be secured, identified, segregated, handled, or accounted for in such a manner as to ensure their safeguard or integrity. Protected cargo is subdivided into classified, controlled, pilferable, and sensitive items.

Protected Harness
A harness that employs some overall outer covering to provide additional mechanical protection for the wires and/or cable contained therein. The added protection may consist of an overbraid, tape wrap, conduit, or some other form of protection.

Prototype
(1) A preproduction, functioning specimen that is the first of its type, typically used for the evaluation of design, performance, and/or production potential. (2) A model suitable for evaluation of design, performance, and production potential.

Prototype Model
A model suitable for complete evaluation of mechanical and electrical form, design, and performance. It is in final mechanical and electrical form, uses approved parts, and is completely representative of final equipment.

Provisions
Space in all feed-through connections and in all wire runs that will allow future incorporation in the aircraft or store without modification other than the addition or changes to connectors, cables, and hardware/software necessary to control the added functions.

Pulse
A transition in the magnitude of a quantity, short in relation to the time span of interest.

Pulse Amplitude
The magnitude of a pulse, measured with respect to a specified reference value. *Note*: for a specific designation, adjectives such as "average," "instantaneous," "peak," "root mean-square," etc. should be used to indicate the particular meaning intended.

Pulsed-Arc Welding
A gas-shielded arc-welding process that extends spray transfer to lower welding currents by superimposing a high current "pulse" onto a lower background welding current.

Pulse Rise Time
The interval between the instant at which the instantaneous amplitude first reaches specified lower and upper limits, namely, 10 and 90% of the peak pulse amplitude, unless otherwise stated.

Punctured Code
A higher-rate code obtained by periodically deleting bits from a lower-rate code.

Purchased Item
An item that is sold or traded in the course of conducting normal business operations, is used regularly by commercial industry, or is a specialized version of a supplier's general product line that he routinely customizes.

Purple Plague
A brittle, gold-aluminum compound formed in the presence of silicon.

Pyroshock
The shock environment imposed on the space vehicle components due to the structural response when the space or launch vehicle pyrotechnic device is ignited. Resultant structural response accelerations resemble the form of superimposed complex decaying sinusoidal wave-forms that decay to a few percent of the maximum acceleration in 5–15 ms.

Q

Qualification
The formal process by which a manufacturer's product is examined for compliance with the requirements of a source control drawing for the purpose of approving the manufacturer as a source of supply.

Qualification Testing
Testing of a purchased item performed prior to procurement action to ensure that the item satisfies the specified requirements.

Qualified Products List (QPL)
A list of products, qualified under the requirements stated in the applicable specification, including appropriate identification and reference data with the name and plant address of the manufacturer. The term *QPL* is substitutable for the term Qualified Products List.

Qualifying Symbol
That portion of a rectangular-shaped logic symbol that denotes its logic function.

Qualitative Information
Information presented by a display in a manner that permits the display user to assess the information without requiring attention to an exact numerical quantity.

Qualitative Maintainability Requirement
A maintainability requirement expressed in qualitative terms; e.g., minimize complexity, design for a minimum number of tools and items of test equipment, and design for optimum accessibility.

Quality Assurance (QA)
A planned and systematic pattern of all actions necessary to provide adequate confidence that management and technical planning and controls are adequate to establish correct technical requirements for design and manufacturing and to create products and services that conform to the established technical requirements.

Quality Assurance Provisions (QAP)
QAPs are the documented requirements, procedures, and criteria necessary for demonstrating that designs conform to user requirements and that materiel and associated services conform to approved designs. In the context of ASME Y14.100, QAP is used to convey a document prepared separate from, but in direct support of, the stated drawing requirements.

Quality Assurance Requirements
The tests and inspections necessary to verify that an end item meets the physical and functional requirements for which it was designed, or verify that a component, part, or subassembly will perform satisfactorily in its intended application.

Quality Assurance Software
Computer programs used to test cases, simulators, and validation and verification tools to certify the quality of operation or test computer programs.

Quality Conformance Inspection (QCI)
Quality conformance inspection is defined as a stress test or series of tests (electrical, environmental, mechanical, and/or combinations thereof) imposed on a sample of the parts and or materials from a lot, for the purpose of lot integrity and performance verification.

Quality Conformance Test Circuitry (Test Coupon)
A portion of a printed board panel that contains a complete set of test coupons that are used to determine the acceptability of the board(s) on the panel.

Quality Control (QC)
A management function whereby control of the quality of raw materials, assemblies, produced materiel, and services is exercised for the purpose of preventing production of defective materiel or providing faulty services.

Quality Engineering Planning List (QEPL)
A cross index of quality engineering documentation to engineering documentation.

Quality Factor
In a reactive circuit, the ratio of the reactance in ohms divided by the resistance in ohms.

Quantitative Expression
The exact quantity of volume linear measurement, weight, or count contained in a UI (5 gallons, 100 feet, 10 pounds, 25 each, etc.).

Quantitative Information
Information presented by a display in a manner that permits the display user to observe or extract a numerical value associated with the information. Quantitative information may be displayed in either digital or analog form.

Quantity per Unit Pack (QUP)
The quantity of items in a unit pack given in the terminology of the definitive unit of issue. When a nondefinitive unit of issue is assigned to the stock item, it may be further quantified by a unit of measure and measurement quantity.

Quartz Clock
A clock containing a quartz oscillator that determines the accuracy and precision of the clock.

Quasi-Analog Signal
A digital signal that has been converted to a form suitable for transmission over a specified analog channel. *Note*: the specification of the analog channel should include frequency range, bandwidth, signal-to-noise ratio, and envelope delay distortion. When quasi-analog form of signaling is used to convey message traffic over dial-up telephone systems, it is often referred to as *voice-data*. A modem may be used for the conversion process.

Quasi-Isotropic Laminate
A laminate approximating isotropy by orientation of plies in several or more directions.

Que or Queue
A collection of items, such as telephone calls, arranged in sequence. *Note*: queues are used to store events occurring at random times and to service them according to a prescribed discipline that may be fixed or adaptive.

Quench Annealing
Annealing an austenitic ferrous alloy by solution heat treatment.

Quench Hardening
Hardening a ferrous alloy by austenitizing and then cooling rapidly enough so that some or all of the austenite transforms to martensite.

Query Language
A type of dialogue in which users compose control entries for displaying specified data from a database.

Queue
A storage mechanism in a multi-user environment that holds jobs or data to be processed within the operation of a computer or program. Most common ones are termed "first-in, first-out" and "last-in, first-out." In software, it is more often called a *stack*. Used in first-in, first-out list algorithms.

Quick-Look Data
Those data provided at the termination of a test, or at some period during the test, on an expedited basis to provide rapid reviews of results.

R

Rack

A floor-standing structure primarily designed for and capable of supporting equipment. A frame upon which one or more units of equipment are mounted.

Rack and Stack

An ATE that relies on system integration of applicable units of test equipment.

Radian

Metric unit of measurement for plane angles.

Radiance

Radiant power, in a given direction, per unit solid angle per unit of projected area of the source, as viewed from the given direction. *Note*: radiance is usually expressed in watts per steradian per square meter.

Radiant Emittance

Radiant power emitted into a full sphere by a unit area of a source, expressed in watts per square meter.

Radiant Energy

Energy that is transferred via electromagnetic waves; i.e., the time integral of radiant power, usually expressed in joules.

Radiant Power

The time rate of flow of radiant energy, expressed in watts.

Radio Frequency (RF) Compatibility

The ability of the various antenna-connected RF receiver and transmitter subsystems contained within a system to function properly without performance degradation caused by antenna-to-antenna coupling between any two subsystems.

Radome

The housing for a radar antenna, essentially transparent to radio frequency.

RAM

Semiconductor based computer memory that stores program code and data in locations that can be accessed in any order.

Random Effect

A common shift in a group of measurements due to a random level change of a usually uncontrollable factor.

Random Error

That part of data variation due to level changes in uncontrolled factors that affect each observation separately and independently.

Random Failure
(1) Any failure whose occurrence is unpredictable in an absolute sense but is predictable only in a probabilistic or statistical sense. (2) Any failure whose exact time of occurrence cannot be predicted.

Random Sample
A sample selected in such a way that each unit of the population has an equal chance of being selected.

Raster
The closely spaced parallel lines produced on a display device. An image is formed by modulating the intensity of the individual pixels. A binary representation, *raster form*, of the pixels can be used to digitally represent an image.

Raster Count
The number of intersections between all addressable horizontal and vertical grid lines in a raster.

Raster Graphics
The presentation or storage of images in raster forms.

Raster Scan
(1) A method of generating or recording the elements of a display image by a line-by-line sweep across the entire display surface, e.g., the generation of a picture on a television screen. (2) A type of antenna motion in which the pencil beam of radar scans a sector in both the horizontal and vertical planes. The scan of the electron beam on a TV screen is a raster.

Raster Unit
The horizontal or vertical distance between two adjacent addressable points on a display surface.

Rawin
A method of winds aloft observation useful in guided missilery. The determination of wind speeds and directions in the atmosphere above the launching facility and the target. It is accomplished by tracking a balloon-borne radar target, a responder, or radiosonde transmitter with either radar or a radio direction finder. Rawin is an acronym composed of *radar* and *wind*.

Read-Only Memory (ROM)
A memory in which data, under normal conditions, can only be read.

Readouts and Displays
Readouts and displays are devices that are designed primarily to convert electrical information into alphanumeric or symbolic presentations. These devices may contain integrated circuitry to function as decoders or drivers.

Real Time
(1) Pertaining to the actual time during which a physical process transpires. (2) Pertaining to the performance of a computation during the actual time that the related physical process transpires, in order that results of the computation can be used in guiding the physical process.

Reassembly
A technician task for replacement of items removed to gain access to facilitate repair and for closing the equipment for return to service.

Rebuild
To restore to a condition comparable to new by disassembling the item to determine the condition of each of its component part and reassembling it using serviceable, rebuilt, or new assemblies, subassemblies, and parts.

Receiver Figure of Merit
The ratio of rms output noise to the response produced by a single hole-electron pair. This dimensionless quantity is useful because it effectively combines a number of component variables.

Recurring Costs
Costs that are incurred for each item changed or for each service or document ordered.

Red Plague
(1) A copper oxide corrosion product formed on silver plate-over-copper at pinholes or breaks in the silver plate. (2) A powdery brown-red corrosion sometimes found on silver-coated conductors and shield braids. It is fungus-like in appearance and will appear at random spots along the length of a conductor or shield. It most often occurs at the point of crossover in a shield or in the interstices of a stranded conductor.

Reduced Inspection
Inspection under a sampling plan using the same quality level as for normal inspection, but requiring a smaller sample for inspection.

Reduction of Area
The difference between the original cross-sectional area of a tension test specimen and the area of its smallest cross section, usually expressed as a percentage of the original area.

Redundancy
The existence of more than one means for accomplishing a given task, where some number of means must fail before there is an overall failure to the system. *Parallel* redundancy applies to systems where both means are working at the same time to accomplish the task and either of the systems is capable of handling the job itself in case of failure of the other system. *Series* or *standby* redundancy applies to a system where there is an alternative means of accomplishing the task, i.e., the standby redundancy is switched in by malfunction sensing device when the primary system fails.

Redundancy Check
(1) A method of verifying that any redundant hardware or software in a communication system is an operational condition. (2) A check that uses one or more extra binary digits or characters attached to data for the detection of errors.

Redundant Design
Alternative or parallel methods of performing a given function that are not necessary for a system operation but are utilized when the primary function fails.

Reengineering
Examination and alteration of an existing system to reconstitute it in a new form, and the subsequent implementation of the new form.

Reference
To invoke associated data by callout on an engineering drawing. Such callouts may be located on the field of the drawing, in the general notes, in the parts, or elsewhere on the drawing.

Reference Circuit
A hypothetical circuit of specified length and configuration with a defined transmission characteristic, primarily used as a reference for measuring the performance of other circuits and as a guide for planning and engineering of circuits and networks.

Reference Data
Information (including dimensions) that does not govern production or inspection operations. Reference data is indicated by enclosing the data in parentheses or by labeling it REF.

Reference Designation
Letters or numbers, or both, used to identify and locate discrete units, portions thereof, and basic parts of a specific set. (A reference designation is not a letter symbol, abbreviation, or functional designation for an item.)

Reference Designation (Basic)
The simplest form of a reference designation consisting only of a class letter portion and a number (namely, without mention of the item within which the reference-designated item is located.)

Reference Designation (Complete)
A reference designation that consists of a basic reference designation and, as prefixes, all the reference designation that apply to the subassemblies or assemblies within which the item is located, including those of the highest level needed to designated the item uniquely.

Reference Designation (Partial)
A reference designation that consists of a basic reference designation and which may include, as prefixes, some but not all of the reference designations that apply to the subassemblies or assemblies within which the item is located.

Reference Dimension
A dimension, usually without tolerance, used for information purposes only. It is considered auxiliary information and does not govern production or inspection operations. A reference dimension is a repeat of a dimension or is derived from other values shown on the drawing or on related drawings.

Reference Documents
(1) Design activity standards, drawings, specifications, or other documents referenced on drawings or lists. (2) Documents referred to in a TDP element that contains information necessary to meet the design disclosure requirements of that TDP element.

Reference Standards
Standards (that is, primary, secondary, and working standards, where appropriate) used in a calibration program. These standards establish the basic accuracy limits for that program.

Reflow Soldering
A process for joining parts by tinning the mating surfaces, placing them together, heating until the solder fuses, and allowing to cool in the joined position.

Regardless of Feature Size (RFS)
(1) The term used to indicate that a geometric tolerance or datum reference applies at any increment of size of the feature within its size tolerance. No longer is there a symbol for RFS. (2) Indicates a geometric tolerance applies at any increment of size of the actual mating envelope of the feature of size.

Regardless of Material Boundary (RMB)
Indicates that a datum feature simulator progress from MMB (maximum material boundary) toward LMB (least material boundary) until it makes maximum contact with the extremities of a feature.

Register Mark
A symbol used as a reference point to maintain registration.

Registration
The degree of conformity of the position of a pattern, or a portion thereof, with its intended position or with that of any other conductor layer of a board.

Registration Number
The number assigned by the Government to an individual unit of a group of items. The number registers Government ownership, responsibility, and accountability (e.g., vehicle registration numbers).

Reinforced Plastic
A plastic with relatively high stiffness or very high-strength fibers embedded in the composition. This improves some mechanical properties over that of the base resin.

Related Views
Two views that are adjacent to the same intermediate view.

Release
The designation by the contractor that a document is complete and suitable for use. *Release* means that the document is subject to the contractor's configuration control procedures.

Release Agent
 - see MOLD RELEASE AGENT.

Released Data
The configuration management controlled version of the data that has been released in accordance with Government CM standards, after contractor internal review and approval. Released data may be provided to the Government for purposes such as design review.

Reliability
(1) The probability of failure-free performance for a specified interval under stated conditions. (2) The probability that an item can perform its intended function for a specified interval under stated conditions. (For nonredundant items, this is equivalent to definition; for redundant items, this is equivalent to definition of mission reliability.)

Reliability Assurance
All actions necessary to provide adequate confidence that material conforms to established reliability requirements.

Remote Tracks
Tracks generated by other than the co-located sensors on which the air defense system relies to acquire targets, e.g., tracks received from interfacing systems.

Removable Assembly
A removable assembly is defined as an assembly that is easily removable from the aircraft. Examples: dynamotor unit, indicator unit, radio receiver, and radio transmitter.

Repair
The process of returning an item to a specified condition, including preparation, fault location, item procurement, fault correction, adjustment and calibration, and final test.

Repairable
Having the capability of being repaired.

Repairable Item
(1) An item that can be restored to perform all of its required functions by corrective maintenance. (2) An item that is interchangeable with another item, but that differs physically from the original item in that the installation of the replacement item requires operations such as drilling, reaming, cutting, filing, or shimming, in addition to the normal applications and methods of attachment.

Repair Parts
See also Spare Parts. Those support items that are an integral part of the end item or system that are coded as nonrepairable.

Replaceable Module
An item that is designed and packaged in a replaceable unit for ready removal and replacement.

Replaceable Unit
Any unit that is designed and packaged to be readily removed and replaced in an equipment system without unnecessary calibration or adjustment.

Replacement Drawing
A replacement drawing is a new original drawing substituted for the previous original drawing of the same drawing number.

Replacement Item
One that is interchangeable with another item, but that differs physically from the original item in that the installation of the replacement item requires operations such as drilling, reaming, cutting, filing, shimming, etc. in addition to the normal application and methods of attachment.

Replacement Schedule
The specified periods when items of operating equipment are to be replaced. *Replacement* means removal of items approaching the end of their maximum useful life, or the time interval specified for item overhaul or rework, and installation of a serviceable item in its place.

Replacing
Substituting one unit for another unit. Usually done to substitute a properly functioning unit for a malfunctioning unit.

Reproducible
A document or copy thereof that is sufficiently translucent to be used as a printing master in a contact printing reproduction process.

Reproduction
The duplication of original copy by any photographic or photomechanical process.

Reproduction Quality
A term used to describe the highest quality microfilm images; images that are capable of being enlarged to original size to make reproducible copies on translucent paper.

Requiring Authority
An activity (Government, contractor, or subcontractor) that levies testability task or subtask performance requirements on another activity (performing activity) through a contract or other document of agreement.

Residual Hypo
Hypo remaining in film or paper after washing. Since residual hypo has a deleterious effect and reduces permanence, careful control must be maintained in processing to ensure that permissible limits are not exceeded.

Residual Hypo Test
A test method, using mercuric chloride, to measure residual thiosulfate content in films.

Residual Stress
A stress present in a metal that is free of external forces or temperature gradients. Usually the result of fabrication processes, it can be tensile or compressive in nature.

Resin
An organic polymer or prepolymer used as a matrix to contain the fibrous reinforcement in a composite material or as an adhesive. This organic matrix may be a thermoset or a thermoplastic and may contain a wide variety of components or additives to influence handleability, processing behavior, and ultimate properties.

Resin Content
The amount of matrix present in a composite, either by percent weight or percent volume.

Resin Recession
The presence of voids between the barrel of the plated-through hole and the wall of the hole, seen in microsections of plated-through holes in boards that have been exposed to high temperatures.

Resin-Rich
A significant thickness of nonreinforced surface-layer resin of the same composition as that within the base material.

Resin Smear
Resin transferred from the base material onto the surface or edge of the conductive pattern normally caused by drilling.

Resin Starvation
Resin starvation is a deficiency of resin in base material that is apparent after lamination by the presence of weave texture.

Resin Starved Area
Area of composite part where the resin has a noncontinuous smooth coverage of the fiber.

Resist
Coating material used to ask or to protect selected areas of a pattern from the action of an etchant, solder, or plating.

Resistance Soldering
A method of soldering in which a current is passed through the soldering area by contact with one or more electrodes, thus heating it.

Resistance Spot Welding (RSW)
A resistance welding process that produces coalescence at the faying surfaces in one spot by the heat obtained from the resistance to electric current through the work parts held together under pressure by electrodes. The size and shape of the individually formed welds are limited primarily by the size and contour of the electrodes.

Resistance Welding
A group of welding processes in which fusion is obtained by the heat obtained from resistance of the work to the flow of electric current in a circuit of which the work is a part, and by the application of pressure.

Response Time
(1) The time a system takes to react to a given input. (2) In a data system, the elapsed time between the end of transmission of an enquiry message and the beginning of the receipt of a response message, measured at the enquiry originating station. (3) The time a functional unit takes to react to a given input.

Restraint
The application of force to a part to simulate its assembly or functional condition resulting in possible distortion of a part from its free-state condition.

Restrictive Markings
Markings on technical data or computer software that limit the Government's right to use, duplicate, or disclose such data or software.

Resubmitted Lot
A lot that has been rejected, subjected to either examination or testing, or both, for the purpose of removing all defective units that may or may not be reworked or replaced and submitted again for acceptance.

Resultant Condition
(1) The variable boundary generated by the collective effects of a size feature's specified maximum material condition of least material condition, the geometric tolerance for the material condition, the size tolerance, and the additional geometric tolerance derived from the feature's departure from its specified material condition. (2) The single worst-case boundary generated by the collective effects of a feature of the size's specified MMC or LMC, the geometric tolerance for that material condition, the size tolerance, and the additional geometric tolerance derived from the feature's departure from its specified material condition.

Retest Okay (RTOK)
(1) A unit under test that malfunctions in a specific manner during operational testing, but performs that specific function satisfactorily at a higher-level maintenance facility. (2) A unit that was identified as malfunctioning in a particular manner at one maintenance level, but in which that specific malfunction could not be duplicated at a higher-level maintenance facility.

Retrofit
The incorporation of new design parts resulting from an approved engineering change to an item's current approved product configuration documentation into already accepted and/or operational items.

Retrograde
To move or appear to move backward. To apply force opposite the direction of flight, usually by firing a rocket on command in the direction of travel to change orbital parameters or to effect reentry.

Reusable Software
Software developed in response to the requirements for one application that can be used, in whole or in part, to satisfy the requirements of another application.

Reverberation Time
The time that would be required for the mean-square sound pressure level, originally in a steady state, to fall 60 dB after the source is stopped.

Reverse Image
The resist pattern on a printed board used to allow for the exposure of conductive areas for subsequent plating.

Reversion
A chemical reaction in which a polymerized material degenerates, at least partially, to a lower polymeric state or the original monomer. It is usually accompanied by significant changes in physical and mechanical properties.

Revision
Any change to an original drawing that requires the revision level to be advanced.

Revision Authorization
A revision authorization is a document such as a Notice of Revision (NOR), Engineering Change Notice, or Revision Directive, that describes the changes to be made to the drawing in detail and is issued by the activity having the authority to revise the drawing.

Rework
A procedure applied to a nonconformance that will completely eliminate it and result in a characteristic that conforms completely to the drawings, specifications, or contract requirements.

Right-Hand Thread
A screw thread that is screwed in or on clockwise.

Rings, Newton's
When two polished surfaces are cleaned and placed in contact with a thin air film between them, reflected beams of light from the two adjacent surfaces interfere to form a series of rings or bands knows as *Newton's rings* or *fringes*. By counting these bands from the point of actual contact, the departure of one surface from the other is determined. The regularity of the fringes maps out the regularity of the distance between the two surfaces.

Ripple
Ripple is the cyclic variation of voltage about the mean level of the voltage during steady-state DC electrical system operation. The ripple voltage may contain multiple frequencies and includes noise not generated by transients. The average value of ripple is zero.

Riser
A reservoir in a mold that supplies molten material for part contraction during solidification.

Risk
An expression of the possibility of a mishap in terms of hazard severity and hazard probability.

Rod
A solid wrought product that is long in relation to its circular cross section, which is not less than 0.375 inch diameter.

Room Temperature Ambient (RTA)
(1) An environmental condition of $73 \pm 5°F$ ($23 \pm 3°C$) at ambient laboratory relative humidity.
(2) A material condition where, immediately following consolidation/cure, the material is stored at $73 \pm 5°F$ and at a maximum relative humidity of 60%.

Root
That surface of the thread that joins the flanks of adjacent thread forms and is immediately adjacent to the cylinder or cone from which the thread projects.

Root Diameter
The diameter of an imaginary cylinder or cone bounding the bottom of the roots of a screw thread. Root diameter is a nonpreferred term for the minor diameter of an external thread or the major diameter of an internal thread.

Root Face
That portion of the groove face adjacent to the root of the joint.

Root (Joint)
That portion of a joint where the members are closest to each other.

Root Mean Square (RMS)
The square root of the average of the squares of the values of a periodic quantity taken throughout one complete period. It is the effective value of a periodic quantity.

Root Opening (GAP)
The separation between the members to be joined, at the root of the joint.

Root Pass
The welding pass made to lay a bead in the root opening; the first pass of a multipass weld.

Roughness
Roughness consists of the finer irregularities of the surface texture, usually including irregularities that result from the inherent action of the production process. These are considered to include traverse feed marks and other irregularities within the limits of the roughness sampling length.

Roundness
   - see CIRCULARITY.

Router
A program that automatically determines the routing path for the component connections on a PC board or hybrid; also may be referenced in connection with the actions of a profiler.

Routing
The placement of interconnections on a PC board.

RSS
A metadata push technology that can identify changes in data and initiate a content push to the end user, without the user having to search it out and pull it from the site.

Ruggedized
(1) Commercial off the shelf (COTS) or modified COTS equipment that is modified to meet specified service requirements. Modified COTS involves modification to meet functional requirements; ruggedized incorporates modification to meet service requirements. This may be in the form of added parts, such as shields and shock mounts, power conditioners, and so

forth, or in the form of direct modification to COTS equipment. (2) Physical and operational characteristics that allow equipment to withstand rough handling and extreme or hostile environments.

Run-In
To operate mechanical items under specified environmental and test conditions to eliminate early failures and to stabilize the items prior to actual use.

Runout
(1) Runout is a composite tolerance used to control the functional relationship of one or more features of a part to a datum axis. The types of features controlled by runout tolerances include those surfaces constructed around a datum axis and those constructed at right angles to a datum axis. (2) As applied to screw threads, refers to circular runout of the major and minor cylinders with respect to the pitch cylinder. Circular runout controls cumulative variations due to eccentricity and out-of-roundness. The amount of runout is usually expressed in terms of full indicator movement (FIM).

Rusting
A type of corrosion attack limited to ferrous materials that results in reddish-brown corrosion products.

S

Sabot

An attachment that fits within the launching tube of one projectile to permit the positioning and firing of a projectile of smaller dimension. The sabot normally is detached from the projectile in flight, although it may incorporate a rocket motor for added velocity.

Safe Working Load

The maximum assigned load the device or equipment can operationally handle and maintain. This value is marked on the device indicating maximum working capacity. This is also the load referred to as "rated load" or "working load limit." If the device has never been down-rated or uprated, this also is the "manufacturer's rated load."

Safety

Freedom from those conditions that can cause death, injury, occupational illness, or damage to or loss of equipment or property.

Safety Critical

A category of subsystems and equipment whose degraded performance could result in loss of life or loss of vehicle or platform.

Sag

To cause a sheet of glass to conform to a ceramic or metal form, by heating the glass to its softening point and allowing it to settle. In the geometric sense, it is also used as an abbreviation for *sagitta*, the height of a curve measured from the chord.

Sample

A sample consists of one or more units of product drawing from a lot or batch, the units of the sample being selected at random without regard to their quality. The number of units of product in the sample is the sample size.

Sample Size

The number of units of product in the sample selected for inspection.

Sample Unit

A unit of product selected to be part of a sample.

Sampling, Biased

Sampling procedures that will not guarantee a truly representative or random sample.

Sampling Frequency

The sampling frequency is the ratio between the number of units of product randomly selected for inspection at an inspection station to the number of units of product passing the inspection station.

Sampling Plan

A sampling plan indicates the number of units of product from each lot or batch that are to be inspected (sample size or series of sample sizes) and the criteria for determining the acceptability of the lot or batch (acceptance and rejection numbers).

Sandwich Construction
A structural panel concept consisting, in its simplest form, of two relatively thin, parallel sheets of structural material bonded to, and separated by, a relatively thick, lightweight core.

Saturation
(1) An equilibrium condition in which the net rate of absorption under prescribed conditions falls essentially to zero. (2) In a communication system, that condition wherein a component of the system has just reached its maximum traffic-handling capacity. (3) That point at which the output of a linear device deviates significantly from being a linear function of the input when the input signal is increased.

Scaling
A formation at high temperature of thick corrosion product layers on a metal surface, or the deposition of water-insoluble constituents on a metal surface.

Scan Rate
The process of examining an area, a region in space, or a portion of the radio frequency spectrum point by point in an ordered sequence per unit of time.

Schematic Diagram
(1) A drawing that shows, by means of graphic symbols, the electrical connections, components, and functions of a specific circuit arrangement. (2) A diagram that shows, by means of graphic symbols, the electrical connections and functions of a specific circuit arrangement. The schematic diagram facilitates tracing the circuit and its functions without regard to the actual physical size, shape, or location of the component device or parts.

Scissor
In computer graphics, to remove parts of display elements that lie outside defined bounds. Synonymous with *clip*. See also Window.

Scrap
Nonconforming material that is not usable for its intended purpose and cannot be economically reworked or repaired in a manner acceptable to the Government.

Scratch
Any marking or tearing of the surface appearing as though it had been done by either a sharp or rough instrument. Scratches occur on sheet glass in all degrees from various accidental causes. *Block reek* is a chain-line scratch produced in polishing. A *runner-cut* is a curved scratch caused by grinding. A *sleek* is a hairline scratch. A *crush* or *rub* is a surface scratch or series of small scratches generally caused by mishandling.

Screenable Latent Defect
A latent defect that is accelerated to failure by a screen and then detected by test.

Screening Experiments
Stress screening applied to preproduction equipment in order to derive data such as screen parameters for planning the overall Environmental Stress Screening program.

Screening Regimen
A combination of stress screens applied to equipment, identified in the order of application (i.e., assembly, unit and system screens).

Screening Strength
The probability that a specific screen will precipitate a latent defect to failure and detect it by test, given that a latent defect susceptible to the screen is present. It is the product of precipitation efficiency and detection efficiency.

Screen Parameters
Parameters that relate to screening strength (e.g., vibration G levels, temperature rate of change, and time duration).

Screen Printing
The transferring of an image to a surface by forcing a suitable media with a squeegee through an imaged screen mesh.

Screw, Cap, Hexagon Head
A bolt, machine, hexagon head except that the length of the unthreaded portion is not controlled.

Screw, Close Tolerance
A bolt, close tolerance except that the head has an internal socket, recess, or slot and the minimum tensile strength may be any value.

Screw, Drive
A hardened cylindrical fastener with multiple spiral flutes on its shank. It also has an end smaller in diameter than the outside diameter of the spiral flutes, which acts as a pilot when driven into a drilled hole.

Screw, Eye
A fastening device with one end formed in the shape of an eye and other end threaded with a lag or wood screw type of thread.

Screw, Instrument
A screw, machine except that the thread diameter is less than 0.060 inch.

Screw, Machine
An externally threaded fastener whose threaded portion is of one nominal diameter. The unthreaded portion has a tolerance greater than that specified for bolt, close tolerance. For threaded sizes 0.060 through 0.164 inch, any head may be used except screw, cap, socket head or bolt, internal wrenching. For thread sizes 0.190 and larger, any recess, slot or socket (except screw, cap, socket head, or bolt, internal wrenching) head may be used.

Screw, Self-Locking
A screw, machine or screw, cap, socket head with the added characteristic of a locking feature incorporated in the design of the head or in the threads.

Screw, Shoulder
A screw, machine except that it has a round unthreaded neck or shank, all or part of which is of greater diameter than the threaded portion.

Screw, Tapping, Thread Cutting
A hardened externally threaded fastener whose thread extends from a tapered end to the bearing surface of the head and is interrupted by flutes or slots to permit cutting its own mating thread.

Screw, Tapping Thread Forming
A hardened externally threaded fastener whose thread usually extends from a gimlet or dog type point to the bearing surface of the head and designed to form its own mating thread.

Screw, Wood
An unhardened externally threaded fastener whose continuous thread extends from a gimlet point for a distance of approximately two thirds of the length of the screw and which is designed to be driven with an inserted driver.

Scribe and Cleave
A technique to prepare fibers for termination in which fibers are lightly scribed then pulled apart to produce cleavage perpendicular to the fiber axis.

Scrim
A low-cost reinforcing fabric made from continuous filament yarn in an open-mesh construction. Used in the processing of tape or other B-stage material to facilitate handling.

Seal Welds
Seal welds are welds provided for a fluid containment function only, as in a closure where strength is provided by a separate device. This definition does not apply to boiler, economizer, and superheater tube-to-header seal welds.

Sector Scan
Motion of the antenna assembly back and forth through a limited azimuth range in contrast to circular (360°) scan or revolution.

Seed
A term used to denote a gaseous inclusion having an extremely small diameter in glass.

Selected Item
A selected item is an existing item, under the control of another design activity or defined by a nationally recognized standardization document, that is subject to refined acceptance criteria (such as fit, tolerance, performance, or reliability) to meet design requirements.

Selected Item Drawing
A selected item drawing defines refined acceptance criteria for an existing item under the control of another design activity or defined by a nationally recognized standard that requires further selection, restriction, or testing for such characteristics as fit, tolerance, material, performance, reliability, etc. within the originally prescribed limits. This drawing type generally permits selection to be performed by any competent inspection or test facility including those of the original manufacturer, the selecting design activity, or a third party.

Self Test
A test or series of tests, performed by a device upon itself, that shows whether the device is operating within designed limits. This includes test programs on computers and automatic test equipment that check out their performance status and readiness.

Semiautomatic Arc Welding
Arc welding with equipment that controls only the filler-metal feed. The advance of welding is manually controlled.

Semiconductor Diode
A semiconductor device having two terminals and exhibiting a nonlinear voltage-current characteristic.

Semper Fidelis (Semper Fi)
Always faithful.

Sensitive Electronic Devices (SEDs)
Electronic parts having highly sensitive characteristics (e.g., thin-layered internal composition) and delicate, miniaturized construction, which are susceptible to damage or degradation, in various degrees, from environmental field forces (electrostatic, electromagnetic, magnetic, or radioactive) as well as more mundane sources such as corrosion, shock, vibration, foreign particle intrusion, biological contamination, thermal stress, and thermal shock. This susceptibility also extends to the standard electronic modules, printed circuit boards, printed wiring boards, and circuit card assemblies containing one or more of these sensitive electronic parts.

Sensitive Items
Items such as small arms, ammunition, and explosives with the potential for use during civil disturbances and domestic unrest, or by criminal elements. In the hands of militant or revolutionary organizations, these items present a definite threat to public safety.

Separable Assembly
Multiple pieces capable of being disassembled and reassembled without damaging or destroying pieces of the assembly.

Separate Parts List
A parts list prepared as a document separate from the engineering drawing with which it is associated, and one that may be revised independently of the drawing.

Sepia
(1) A yellow-brown pigment. (2) A term used to describe a photographic that has been chemically treated to render the tones sepia. (3) A term applied to a brown-colored photograph or diazo-type reproduction.

Sequential Electrochemical Reduction Analysis (SERA)
A chronopotentiometric reduction method for assessing tin-lead finish solderability.

Sequential Laminating Process
A process for making multilayer printed wiring boards by laminating increments of double-sided boards together. The circuitry layers are interconnected with via holes and through connections. The terms *sequential laminating process, conventional laminating process*, and *sequential plating process* are not synonymous. The distinction between these processes is the structure of the multilayer boards produced. The *sequential laminating process* is used to make multilayer boards, consisting of increments of double-sided boards, using via hole interfacial connections between the two circuitry layers on opposite sides of each individual increment. The *sequential plating process* is used to make multilayer boards by using the electroplating process to build up each circuitry layer, one at a time.

Sequential Lamination
A type of multilayer printed wiring board, consisting of several individual single- or double-sided boards laminated together. Each double-sided increment usually contains via holes, and the entire laminated assembly usually contains plated-through holes, which are through connections. *Sequential lamination* is a final assembly, with plated-through holes, composed of individual multilayer or double-sided boards, which may contain plated-through holes.

Sequential Logic Function
A logic function wherein, for at least one combination of states of the input or inputs, there exists more than one state of the output or outputs. The outputs are functions of variables in addition to the input, such as time, previous internal states of the element, etc.

Serial Number
(1) The number on the item assigned by the manufacturer or the Government for identification or control. (2) The unique notation that identifies a single unit of a family of like units, normally assigned sequentially. The identifier *SERNO* may be used to avoid confusing with other identifiers and when marking space allows. *Note*: characters are normally numeric or alphanumeric, with special characters as allowed by established standards.

Serviceability
The design, configuration, and installation features that will minimize periodic or preventive maintenance requirements, including the use of special tools, support equipment, skills, and manpower, and enhance the ease of performance of such maintenance, including inspection and servicing.

Service Requirements
Parameters related to the ability of equipment to perform in its application, including but not limited to environmental conditions, auxiliary support services, and equipment supportability.

Set
(1) A unit or units and necessary assemblies, subassemblies, and parts connected or associated together to perform an operational function. (*Set* is also used denote a collection of like parts such as a tool set or a set of tires.) (2) The permanent distortion from the manufactured dimension that occurs when a spring is stressed beyond the elastic limit of the material. (3) The strain remaining after complete release of the force producing the deformation.

Setscrew
An externally threaded device whose threaded portion is one of nominal diameter with or without a head and having a cup, cone or other type of machined point designed to prevent or restrict relative movement of parts and designed to be driven with either a wrench or inserted driver.

Shadowing
A conditioning occurring during etchback in which the dielectric material, immediately next to the foil, is incompletely removed although acceptable etchback may have been achieved elsewhere.

Shall
Establishes a mandatory requirement.

Sheet
A rolled product that is rectangular in cross section, with thickness 0.006 inch and less than 0.250 inch and with slit, sheared or sawed edges. There is an overlap in the thickness range of 0.006–0.0079 inch defined for foil and sheet. Sheet products in this gage range are supplied to sheet product specifications.

Shelf Life
(1) The length of time a material, substance, product, or reagent can be stored under specified environmental conditions and continue to meet all applicable specification requirements and/ or remain suitable for its intended function. (2) Total period of time beginning with date manufactured, date cured, date assembled, and date packed, that an item may remain in the combined wholesale and retail storage system and still be suitable for issue and/or use by the user.

Shelf-Life Code (SLC)

A code assigned to a shelf-life item to identify the period of time beginning with the date of manufacture, date of cure (for elastomeric and rubber products only), date of assembly, or date of pack (subsistence only), and ending with the date by which the item must be used (expiration date) or subjected to inspection, test, restoration, or disposal action.

Shelf Life Item

An item of supply that possesses deteriorative or unstable characteristics to the degree that a storage time period must be assigned to ensure that the item will perform satisfactorily in service.

Shelf-Life Item Type I

An individual item of supply which is determined through an evaluation of technical test data and/or actual experience, to be an item with a definite nonextendible period of shelf-life. One exception is Type I medical shelf-life items, that may be extended if they have been accepted into and passed testing for extension in the DoD/Federal Drug Administration (FDA) Shelf-Life Extension Program (SLEP).

Shelf-Life Item Type II

An individual item of supply having an assigned shelf-life time period that may be extended after completion of visual inspection/certified laboratory test, and/or restorative action. action.

Shield

A housing, screen, or cover that substantially reduces the coupling of electric and magnetic fields into or out of circuits or prevents the accidental contact of objects or persons with parts or components operating at hazardous voltage levels.

Shielded Metal-Arc Welding

An arc-welding process in which fusion is obtained by heating with an electric arc between a covered metal electrode and the work. Decomposition of the electrode provides shielding. Pressure is not used, and filler metal is obtained from the electrode.

Shielding Effectiveness

A measure of the reduction or attenuation in the electromagnetic field strength at a point in space caused by the insertion of a shield between the source and that point.

Shielding, Electronic

A physical barrier, usually electrically conductive, designed to reduce the interaction of electric or magnetic fields upon devices, circuits, or portions of circuits.

Shipping Container

An exterior container that meets carrier regulations and is of sufficient strength, by reason of material, design, and construction, to be shipped safely without further packing (e.g., wooden boxes or crates, fiber and metal drums, and corrugated and solid fiberboard boxes).

Should

Should and *may* are used when it is necessary to express nonmandatory provisions.

Sidebands

Two bands of frequencies, on either side of the carrier frequency of a modulated radio signal including carrier frequency of a modulated radio signal including components whose frequencies are, respectively, the sum and difference of the carrier and the modulating frequencies.

Side Lobe
A portion of the radiation pattern from an antenna, other than the main lobe, and usually much smaller in strength.

Sight, Line of (LOS)
The line of vision; the "optical axis" of a telescope or other observation instrument. The straight line connecting the observer with the aiming point; the line along which the sights are set.

Signal
An electrical impulse of a predetermined voltage, current, polarity, and pulse width.

Signal Conductor
An individual conductor used to transmit an impressed signal.

Signal Plane
A conductor layer intended to carry signals rather than serve as a ground or perform another fixed voltage function.

Signal Return
A current-carrying path between a load and the signal source. It is the low side of the closed-loop energy transfer circuit between a source-load pair.

Significant Digit
Any digit that is necessary to define a value or quantity.

Silver Halide
A compound of silver and one of the following elements known as halogens: chlorine, bromine, Iodine, fluorine.

Single Hop
A connection between two terminals requiring the use of a single satellite.

Single-Line (One-Line) Diagram
A diagram that shows, by means of single lines and graphic symbols, the course of an electric circuit or system of circuits and the component devices or parts used therein.

Single-Sided Board
A printed board with a conductive pattern on one side only.

Single Start Thread
A screw thread having the lead equal to the pitch.

Size, Actual
(1) The general term for the size of a produced feature. This term includes the actual mating size and the actual local sizes. (2) The measured value of any individual distance at any cross section of a feature of size.

Size, Actual Local
The value of any individual distance at any cross section of a feature.

Size, Limits of
The specified maximum and minimum sizes.

Size, Nominal
The designation used for purposes of general identification.

Size, Resultant Condition
The actual value of the resultant condition boundary.

Size, Virtual Condition
The actual value of the virtual condition boundary.

Skew
(1) In parallel transmission, the difference in arrival time of bits transmitted at the same time. (2) For data recorded on multichannel magnetic tape, the difference in time of reading bits recorded as a single line. (3) In facsimile systems, the angular deviation of the received frame from rectangularity due to asynchronism between scanner and recorder.

Sleek
A polishing scratch without visible conchoidal fracturing of the edges.

Sleeper
Wood member nailed to floor and butted against the lading to prevent lateral movement.

Sliding Window
A moving window, typically of fixed duration, that defines the data to be analyzed.

Sliver
A sliver is a slender metallic projection that has been separated from the edge of a printed-circuit conductor.

Slot Time
A time interval, typically of fixed duration, that defines the time structure of a time-shared channel.

Soft Conversion
The process of changing inch-pound measurements to equivalent metric units within the acceptable measurement tolerances without changing the physical configuration. In other words, it is the same both before and after conversion.

Software/Computer Software
(1) A set of computer programs, procedures, and associated documentation concerned with the operation of a data processing system; e.g., compilers, library routines, manuals, and circuit diagrams. (2) Any set of machine readable instructions (most often in the form of a computer program) that directs a computer's processor to perform specific operations. The term is used to contrast with computer hardware, the physical objects (processor and related devices) that carry out the instructions. Hardware and software require each other; neither has any value without the other.

Software Development File (SDF)
A repository for collection of material pertinent to the development or support of software. Contents typically include (either directly or by reference) design considerations and constraints, design documentation and data, schedule and status information, test requirements, test cases, test procedures, and test results.

Software Development Library (SDL)
A controlled collection of software, documentation, and associated tools and procedures used to facilitate the orderly development and subsequent support of software. The SDL includes the Development Configuration as part of its contents. A software development library provides storage of and controlled access to software documentation in human-readable form, machine-readable form, or both. The library may also contain management data pertinent to the software development project.

Software Drawing
A software drawing describes the characteristics of the software and its master media, used for programming each applicable device or assembly.

Software Engineering Environment
The set of automated tools, firmware devices, and hardware necessary to perform the software engineering effort. The automated tools may include, but are not limited to, compilers, assemblers, linkers, loaders, operating system, debuggers, simulators, emulators, test tools, documentation tools, and database management systems.

Software Plans
A collective term used to describe the contractor's plans, procedures, and standards for software management, software engineering, software qualification, software product evaluation, and software configuration management.

Software Quality
The ability of a software product to satisfy its specified requirements.

Software Requirements Specifications (SRS)
A software requirements specification is a document that specifies the detailed requirements (functional, interface, performance, qualification, etc.) allocated to a particular computer software configuration item.

Software Support
The sum of all activities in place to ensure that implemented and fielded software continues to fully support the operational mission of the software.

Software Test Description (STD)
A software test description is a document that identifies the input data, expected output data, and evaluation criteria that compose the test cases. The STD also contains the necessary procedures to perform the formal testing of a computer software configuration item.

Software Test Environment
A set of automated tools, firmware devices, and hardware necessary to test software. The automated tools may include, but are not limited to, test tools such as simulation software, code analyzers, etc., and may also include those tools used in the software engineering environment.

Software Test Plan
A software test plan is a document that describes the formal qualification test plans for one or more computer software configuration items, identifies the software test environment resources required, and provides schedules for the activities. In addition, the software test plan identifies the individual tests that are to be performed.

Software Unit
Any logical set or grouping of instructions to a computer, such as a module or package. Contrast with *program*.

Solderability
(1) The capacity of a metal to be soldered under the fabrication conditions imposed upon a specific, suitably designed structure. (2) The property of a metal to be wetted by solder.

Soldering
(1) A joining process in which a bond is obtained by filling the joint with fusible alloys of relatively low melting temperature. (2) A group of joining processes wherein coalescence is

produced by heating to a suitable temperature and by using a filler metal having a liquidus not exceeding 800°F and below the solidus of the base metals.

Solderless Wrap
A method of connecting a solid wire to a square, rectangular, or V-shaped terminal by tightly wrapping the wire around the terminal with a special tool.

Solder Mask
 - see RESIST.

Solder Paste
Finely divided particles of solder, with additives to promote wetting and to control viscosity, tackiness, slumping, drying rate, etc., that is suspended in a paste flux.

Solder Paste Flux
Solder paste without the solder particles.

Solder Plugs
Cores of solder in the plated-through holes of a printed board.

Solder Projection
An undesirable protrusion of solder from a solidified solder joint or coating.

Solder Resist
 - see RESIST.

Solder Side
The side of a printed board that is opposite to the component side.

Solid Armor
Solid armor is all basic armor materials, composites, and stacked arrangements having no air space between plate elements.

Solids Modeling System
An interactive computer graphics systems used to electronically represent complete three-dimensional shapes of mechanical parts and assemblies.

Solution Heat Treatment
Heating an alloy to a suitable temperature, holding at the temperature long enough to allow one or more constituents to enter into solid solution, and then cooling rapidly enough to hold the constituents in solution. The alloy is left in a supersaturated, unstable state and may subsequently exhibit *quench aging*.

Source
In communications, that part of a system from which messages are considered to originate.

Source Control Drawing
A source control drawing depicts an existing commercial or vendor item that exclusively provides the performance, installation, and interchangeable characteristics required for one or more specific critical applications. A source control drawing provides an engineering description and acceptance criteria for purchased items that require design activity imposed qualification testing and exclusively provides performance, installation, and interchangeability specific characteristics required for critical application. It includes a list of approved manufacturers, the manufacturer's item identification, and acceptance criteria for items that are interchangeable in specific applications. The source control drawing establishes item

identification for the controlled item. The approved items and sources listed on a source control drawing are the only acceptable items and sources.

Source Data
Documentation developed by a contractor to support equipment items developed by that contractor. Source data may stand alone or may be incorporated into other documentation when the hardware is integrated into, attached, or otherwise becomes a part of other equipment.

Source Documents
User's documents that are a source if data eventually processed by the computer program, such as target lists, supply codes, parts lists, maintenance forms, bills of lading, etc.

Source Efficiency
In optical systems, the ratio of emitted optical power of a source to the input electrical power.

Source Language
A language from which statements are translated.

Source Program
A computer program expressed in a programming language.

Space, Aerospace Airborne
"Airborne" denotes those applications peculiar to aircraft and missile or other systems designed for operation primarily within the earth's atmosphere; "space" denotes application peculiar to spacecraft and systems designed for operation near or beyond the upper reaches of the earth's atmosphere; and "aerospace" includes both airborne and space applications.

Spaced Armor
Spaced armor is an armor arrangement consisting of two or more individual plate elements where each plate is spaced from the adjoining one. Spaced armor is especially effective as a protection system against shaped-charge effects.

Spalling
Spalling results when a layer of plating (or armor) material in the area surrounding a warhead impact location is detached or delaminated from the rear face of the plating.

Spare and Repair Parts
(1) Spares are components or assemblies used for maintenance replacement purposes in major end items of equipment. Repair parts are those "bits and pieces," e.g., individual parts or nonrepairable assemblies, required for the repair of spares or major end items. (2) The term *spare parts* includes assemblies, subassemblies, components, parts, repair kits, and raw materials required or anticipated to be required as replacement of like items during manufacturing, operation, maintenance, repair, or overhaul of Aeronautical Vehicle Equipment; Aerospace Ground Equipment, Test Support Equipment, or any other equipment identified as part of the contract including Government furnished property.

Special Characteristics
The pertinent rating, operating characteristics, and other information necessary for identification of the item.

Special Inspection Equipment (SIE)
Either single or multipurpose integrated test units engineered, designed, fabricated or modified to perform special purpose testing of an item. It consists of items or assemblies of

equipment that are interconnected and interdependent so as to become a new functional entity for testing purposes.

Specialized Segment of Industry
A business entity having recognized expertise in developing, manufacturing, or both, specific products or product lines to meet customer requirements.

Special Tooling
Unique tooling that is mandatory to the manufacture of an item. It differs from tooling designed to increase manufacturing efficiency in that the use of the special tool imparts some characteristic to the item that is necessary for satisfactory performance and cannot be duplicated through other generally available manufacturing methods.

Special Tools
Tools not listed in the Federal Supply Catalog.

Specification
A document prepared specifically to support acquisition that clearly and accurately describes essential technical requirements for purchasing materiel. Procedures necessary to determine that the requirements for the materiel covered by the specification have been met are also included.

Specification Change Notice (SCN)
A document used to propose, transmit, and record changes to a specification.

Specification Control Drawing
 - see VENDOR ITEM DRAWING.

Specific Gravity
The ration of the weight of any volume of a substance to the weight of an equal volume of another substance taken as standard at a constant or stated temperature.

Speech Signal Processing
The modification of the electrical signal representing speech to enhance the capability of a speech communications channel. Some examples are simple analog processing, automatic gain control, frequency shaping, peak clipping, and syllabic compression.

Speech Spectrum
A segment of the range of audible frequencies containing the sounds of speech; defined as approximately the range from 80 to 8,000 Hz.

Spin Welding
A process of fusing two objects by forcing them together while one of the pair is spinning until frictional heat melts the interface. Spinning is then stopped and pressure maintained until they are frozen together.

Splay
Splay is the tendency of a rotating drill bit to drill off-center, out-of-round, nonperpendicular holes.

Splice
(1) To join, permanently, physical media that conduct or transmit power or a communication signal. (2) A device that so joins conducting or transmitting media. (3) The completed joint. The proper abbreviation is SPL.

Spot Weld
A weld made between or upon overlapping members wherein coalescence may start and occur on the faying surfaces or may have proceeded from the surface of one member. The weld cross section is approximately circular.

Spot Welding
A welding process in which fusion of the faying surfaces is obtained at one spot. This may be accomplished by either resistance or arc-spot welding techniques.

Spring Rate
The force required to deflect a compression or extension spring one unit of length, or the force required to deflect a torsion spring one degree or one revolution.

Spurious Signals
Undesired voltages in conductors caused by electromagnetic or electrostatic coupling from other conductors or from external sources such as a transformer.

Stabilized Steel
A stainless steel that has been alloyed with a carbide forming metal (e.g., Ti) to reduce or remove susceptibility to weld decay.

Stable Degraded Faults
Faults that will permit indefinite continued use of the equipment at a reduced capability.

Stagnation Pressure
Stagnation pressure is the total pressure sum of the incident overpressure and dynamic wind pressure of a blast wave in the moving air when the air is brought to rest upon impacting an object or structure.

Standard
A document that establishes engineering and technical requirements for items, equipment, processes, procedures, practices, and methods that have been adopted as standard. Standards may also establish requirements for selection, application, and design criteria for materiel.

Standard, Company
A company document that establishes engineering and technical limitations and applications for items, materials, processes, methods, designs, and engineering practices unique to that company.

Standard Generalized Markup Language (SGML)
A standard that defines a language for document representation that formalizes markup and frees it of system and processing dependencies. It provides a coherent and unambiguous syntax for describing whatever a user chooses to identify within a document.

Standard Hardware Acquisition Reliability Program (SHARP)
A coordinated program, residing at the NWSCC, for providing standard hardware for improved acquisition and reliability.

Standard Microcircuit Drawing
A Government unique drawing type used to define the physical and performance characteristics of commercial microcircuits used in military applications.

Standardization
The use of common items, parts, materials, and practices through the life cycle of systems and equipment.

Standardization Document
A document developed by the Government or private sector association, organization, or technical society that plans, develops, establishes, or coordinates standards, specifications, handbooks, or similar documents for the purpose of standardizing items, materials, processes, or procedures.

Standardized Military Drawing (SMD)
An SMD is a control drawing, and it shall disclose the applicable configuration, envelope dimensions, mounting and mating dimensions, interface dimensional characteristics, specified performance requirements, nuclear effects, and inspection and acceptance test requirements for microcircuits in a military application. SMDs depict the Government's requirements for an existing commercial item tested for a military application, disclosing applicable configuration, envelope dimensions, mounting and mating dimensions, interface dimensional characteristics, specified performance requirements, and inspection and acceptance test requirements as appropriate for a military environment.

Standard Measuring Equipment
Commercially available measuring equipment commonly used in a machine shop.

Standard, Nongovernment
A standardization document developed by a private-sector association organization or technical society that plans, develops, establishes, or coordinates standards, specifications, handbooks, or related documents.

Standard Tools
Standard tools (normally hand tools) used for the assembly, disassembly, inspection, servicing, repair, and maintenance of equipment, and which are manufactured by two or more recognized tool manufacturing companies and listed in those companies' catalogs.

Standing Wave
(1) Distribution of current and voltage on a transmission line formed by two sets of waves traveling in opposite directions, and characterized by the presence of a number of points of successive maxima and minima in the distribution curves. (2) Sinusoidal distribution of current and voltage amplitudes along a transmission line as a result of the reflection of energy from a point where a mismatch of impedance occurs. Also called *stationary wave.*

Start Part or Template
An initial 3D solid model containing basic model formatting information such as attributes, parameters, datum planes, notes, etc. required in each 3D solid model.

Stastical Tolerancing (ST)
In geometric dimensioning and tolerancing it is a symbolic means of indicating that a tolerance is based on statistical tolerancing. If the tolerance is a statistical geometric tolerance, the symbol shall be placed in the feature control frame following the stated tolerance and any modifier. If the tolerance is a statistical size tolerance, the symbol shall be placed adjacent to the size dimensioning.

Static Electricity
The stationary electrical charge produced and accumulated or stored on the surface of materials due to triboelectric action (charge generation by friction, such as airflow, or by adhesive forces during separation), particle impingement, or electromagnetic field inducement.

Step-and-Repeat
A method by which successive exposures of a single image are made to produce a multiple image production master.

Storage Time
Time during which a system or equipment is presumed to be in operable condition but is being held for subsequent use.

Store
Any device intended for internal or external carriage and mounted on an aircraft suspension and release equipment, whether or not the item is intended to be separated in flight from the aircraft. Stores are classified in two categories: *Carriage Stores* and *Mission Stores*.

Straightness
Straightness is a condition in which an element of a surface or an axis is a straight line.

Stress-Accelerated Corrosion
A form of intergranular attack that can occur in the absence of stress but is accelerated by the application of a tensile stress, which opens up any cracks and permits the corrodent to attack the metal at the crack tips.

Stress Adjustment Factor
The ratio of the incoming defect density at the anticipated field stress level to the incoming defect density at the baseline stress level.

Stress Corrosion
A specific type of accelerated corrosion resulting from the combined effects of mechanical tensile stress and corrosive environments.

Stress-Corrosion Cracking
A premature failure of a metal as the result of the combined action of tensile stresses and a corrosive environment. The surface tensile stresses may be residual or applied.

Stress Relieving
(1) Uniform heating of a structure or portion thereof to a sufficient temperature, below the critical range, to relieve the major portion of the residual stresses, followed by uniform cooling. (2) To remove residual stresses caused by the forming operation by applying low-temperature heat treatment after coiling or bending. Also called *strain relieve, stress equalizing, tempering, bluing*, and *baking*.

Stress Screening
The process of applying mechanical, electrical and/or thermal stresses to an equipment item for the purpose of precipitating latent part and workmanship defects to early failure.

Stria
A defect in optical glass consisting of a sharply defined streak of transparent material having a slightly different index of refraction than the body of the glass.

Striae
Internal imperfections of glass appearing as wavy distortion.

String
In the user's context, a word, phrase, or number (string of characters) in the test or file. Normally employed in the context of causing the computer to search for, find, or replace a desired "string."

Stripline
A type of transmission line configuration which consists of a single narrow conductor parallel and equidistant to two parallel ground planes.

Structure
Any fixed or transportable building, shelter, tower, or mast that is intended to house electrical or electronic equipment or otherwise support or function as an integral element of an electronics complex.

Stud Welding
An arc-welding process for joining metal studs, or similar parts to another workpiece. Fusion is produced by drawing an electric arc between the parts as they are brought together under pressure.

Subassembly
Two or more parts that form a portion of an assembly or a unit replaceable as a whole, but having a part or parts which are individually replaceable.

Submitted Data
(1) The configuration management controlled master version of the data "formally" submitted to the Government. (2) Data formally provided to the Government in accordance with contractual requirements. Data made available for access is submitted data.

Substitute Item
An item that possesses such functional and physical characteristics as to be capable of being exchanged for another only under specified conditions or in particular applications and without alteration of the items themselves or of adjoining items.

Substrate
- see BASE MATERIAL.

Subsystem
(1) A subsystem is an assembly of two or more components, including the supporting structure to which they are mounted, and any interconnecting cables or tubing. A subsystem is composed of functionally related components that perform one or more prescribed functions. (2) A combination of sets, groups, etc., which performs an operational function within a system and is a major subdivision of the system. An example is an Intercept- Aerial Guided Missile Subsystem.

Subsystem/Equipment
Any electrical, electronic, or electromechanical device, or collection of items, intended to operate as an individual unit and perform a specific set of functions.

Summation Check
A check based on the formation of the sum of the digits of a numeral. The sum of the individual digits is usually compared with a previously computed value.

Superseded (Drawing)
A notation used to indicate that a document has been replaced by another document with a different document number or to indicate that an original has been replaced by a new original.

Supplier
See also Vendor. The party that produces, provides, or furnishes an item and warrants item compliance with the part numbered design drawing specifications and warrants the uniqueness of the part number within the enterprise.

Support Data
Those data items designed to document the logistics support planning and provisioning process. This includes, for example: supply and general maintenance plans and reports, transportation, handling, packaging information, etc., and data to support the provisioning process.

Support Equipment
Items necessary for the maintenance or operation of the system that are not physically part
of the system.

Supporting Structures, Electrical
Normally non-electrified conductive structural elements near to energized electrical conductors
such that a reasonable possibility exists of accidental contact with the energized conductor.

Surfactant
A surfactant is a surface wetting agent which reduces interfacial tension between a liquid and
a solid.

Surface Tension
A property of liquids in which the exposed surface tends to contract to the smallest possible
area, as in the spheroidal information of drops.

Surface Texture
Surface texture is repetitive or random deviations from the nominal surface that form the
three-dimensional topography of the surface. Surface texture includes roughness, waviness,
lay, and flaws.

Surfacing Mat
A thin mat of fine fibers used primarily to produce a smooth surface on an organic matrix
composite.

Survive
The ability of an equipment, subsystem, or system to resume functioning without evidence of
degradation following temporary exposure to an adverse electromagnetic environment. This
implies that the system performance will be degraded during exposure to the environment,
but the system will not experience any damage, such as component burnout, that prevents it
from operating when the levels are removed.

Suspense File
A temporary collection of data saved by the computer for later use.

Suspension and Release Equipment
All airborne devices used for carriage, suspension, employment, and jettison of stores, such
as, but not limited to, racks, adapters, launchers, and pylons.

Sweep, Driven
Sweep triggered only by an incoming signal or trigger.

Sweep, Free-Running
Sweep triggered continuously by an internal trigger generator.

Switch
A device for making, breaking, or changing the connections in an electric circuit.

Symbolic Logic
The discipline in which valid arguments and operations are dealt with using an artificial language designed to avoid the ambiguities and logical inadequacies of natural languages.

Symmetrical Laminate
A composite laminate in which the sequence of plies below the laminate midplane is a mirror
image of the stacking sequence above the midplane.

Symmetrically Opposite Parts
Symmetrically opposite parts are those parts that are mirror images of each other.

Symmetrical Pair
A balanced transmission line in a multi-pair cable having equal conductor resistances per unit length, equal impedances from each conductor to earth, and equal impedances to other lines.

System (Electrical-Electronic)
A combination of two or more sets, which may be physically separated when in operation, and such other assemblies, subassemblies, and parts necessary to perform an operational function or functions.

System (General)
(1) A composite of equipment, skills, and techniques capable of performing or supporting an operational role or both. A complete system includes all equipment, related facilities, material, software, services, and personnel required for its operation and support to the degree that it can be considered a self-sufficient unit in its intended operational environment. (2) A specific suite of computer hardware and software. As used in the terms *Source System* and *Destination System*, the term does not necessarily correspond one to one with *site* or *base* in that most prime contractor sites and DOD installations have more than one system. (3) A composite, at any level of complexity, of personnel, procedures, materials, tools, equipment, facilities, and software. The elements of this composite entity are used together in the intended operational or support environment to perform a given task or achieve a specific production, support, or mission requirement. (4) A combination of sets, groups, etc., which performs an operational function within a system and is a major subdivision of the system. An example is an Intercept-Aerial Guided Missile Subsystem.

System Design Review (SDR)
This review shall be conducted to evaluate the optimization, correlation, completeness, and risks associated with the allocated technical requirements. Also included is a summary review of the system engineering process that produced the allocated technical requirements and of the engineering planning for the next phase of effort. Basic manufacturing considerations will be reviewed, and planning for production engineering in subsequent phases will be addressed. This review will be conducted when the system definition effort has proceeded to the point where system characteristics are defined and the configuration items are identified.

System Engineering
The application of scientific and engineering knowledge to the planning, design, construction, and evaluation of man/machine systems and components. It includes the overall consideration of possible methods for accomplishing a desired result, determination of technical specification, identification and solution of interfaces among parts of the system, development of coordinated test programs, assessment of data, integrated logistic support planning, and supervision of design work.

System/Equipment
A group of units interconnected or assembled to perform some overall electronic function (e.g., electronic flight control system, communications system).

System Ground Point (SGP)
The SGP is a single point on the vehicle structure to which the negative or the neutral of the electrical power system is connected. Electrically isolated power systems may have separate system ground points on the vehicle structure.

System Integrated Test (SIT)

An integral capability of the mission equipment that provides an onboard, automated capability to detect, diagnose, or isolate system failures. The fault detection/isolation capability is used for momentary or continuous monitoring of a system's operational health and for observation/diagnosis as a prelude to maintenance action. System integrated test may be designed as an analysis tool for the overall system, integrated with several subsystems, or designed as an integral part of each removable component.

System Requirements Review (SRR)

The objective of this review is to ascertain the adequacy of the contractor's efforts in defining system requirements. It will be conducted when a significant portion of the system functional requirements has been established.

System Safety

The application of engineering and management principles, criteria, and techniques to optimize safety within the constraints of operational effectiveness, time, and cost throughout all phases of the system life cycle.

System Safety Engineer

An engineer who is qualified by training and/or experience to perform system engineering tasks.

System Safety Engineering

An engineering discipline requiring specialized professional knowledge and skills in applying scientific and engineering principles, criteria, and techniques to identify and eliminate hazards, or reduce the risk associated with hazards.

System Safety Management

An element of management that defines the system safety program requirements and ensures the planning, implementation, and accomplishment of system safety tasks and activities consistent with the overall program requirements.

System Safety Program

The combined tasks and activities of system safety management and system safety engineering that enhance operational effectiveness by satisfying the system safety requirements in a timely, cost-effective manner through all phases of the system life cycle.

System Safety Program Plan

A description of the planned methods to be used by the contractor to implement the tailored requirements, including organizational responsibilities, resources, methods of accomplishment milestones, depth of effort, and integration with other program engineering, and management activities and related systems.

System Segment

A system segment is a major subtier element of a system that is so identified by the responsible program office, either for management expediency or to facilitate separate procurements.

System Specification

A type of program-unique specification that describes the requirements and verification of the requirements for a combination of elements that must function together to produce the capabilities required to fulfill a mission need, including hardware, equipment, software, or any combination thereof.

T

Tack Weld
A weld made to hold parts of a weldment in proper alignment until the final welds are made.

Tailoring
The process of evaluating individual potential requirements, of both streamlined and non-streamlined documents, to determine their pertinence and cost effectiveness for a specific system or equipment acquisition; and modifying these requirements to ensure that each contributes to an optimal balance between need and cost. The tailoring of data requirements consists of determining the essentiality of potential CDRL items and shall be limited to the exclusion of information requirements provisions.

Tangent Plane (T)
In geometric dimensioning and tolerancing it is the symbolic means of indicating a tangent plane. The symbol shall be placed in the feature control frame following the stated tolerance.

Taped Dump
The method by which digital or analog data is stored for a period on a vehicle and then transmitted, or *dumped*, to a ground receiving station.

Taper Thread
A screw thread projecting from a conical surface.

Tape Set
A group of one or more magnetic tapes that collectively represent the collection of related files composing a specific delivery of a document or documents.

Tape Volume
A single reel of magnetic tape with recorded data.

Target Acquisition
The detection and location of a target in relation to a known control point or datum with sufficient accuracy and detail to permit the effective employment of appropriate weapons by the commander.

Target Discrimination
The capability of a system that enables it to distinguish a target from its background, between two or more targets in close proximity, or between targets and decoys.

Target Identification
The act of determining the nature of a target, including whether it is friend or foe.

Target Profile Area
A sectional area of a target as it affects detection, radar reflection, and vulnerability.

Tarnish
A surface discoloration of a metal caused by the formation of a thin film of corrosion product.

Technical Data

(1) Technical data is recorded information (regardless of the form or method of recording) of a scientific or technical nature (including computer software documentation) relating to supplies procured by an agency. Technical data does not include computer software or financial information incidental to contract administration. (2) Recorded information, regardless of form or characteristic, of a scientific or technical nature. It may, for example, document research, experimental, developmental, or engineering work; or be usable or used to define a design or process or to procure, produce, support, maintain, or operate materiel. Technical data includes research and engineering data, engineering drawings and associated lists, specifications, standards, process sheets, technical manuals, technical orders, technical reports, catalog item identifications and related information, and computer software documentation. Technical data does not include computer software or financial, administrative, cost and pricing, and management data, or other information incidental to contract administration.

Technical Data Package (TDP)

(1) A collection of product-related engineering data comprising the EDP and non-EDP data related to the design and manufacture of the item or system. The EDP contains all the descriptive documentation needed to ensure the competitive reprocurement of an item or system. The non-EDP consists of data such as system and development specifications, product specifications, concurrent repair parts list, packaging data sheets, special production tool data, acceptance inspection equipment data, military specifications and standards, repair manuals, supplementary quality assurance provisions, preparation for delivery requirements, and other data as required. (2) A technical description of an item adequate for supporting an acquisition strategy, production, engineering, and logistics support. The description defines the required design configuration and procedures required to ensure adequacy of item performance. It consists of all applicable technical data such as drawings and associated lists, specifications, standards, performance requirements, quality assurance provisions, and packaging details.

Technical Data Package Data Management Product

A data product that is used to monitor and control the development and maintenance of the TDP. A TDP data management product contains information about the TDP rather than the item being documented.

Technical Data Package Document

A document that is part of a TDP element.

Technical Data Package Element

A data product that is an actual component of the TDP. A TDP element provides all or part of the information necessary to define the item being documented by the TDP.

Technical Data Package Data Management Product

A data product that is used to monitor and control the development and maintenance of the TDP. A TDP data management product contains information about the TDP rather than the item being documented.

Technical Manual Specification

A specification used to acquire technical manuals for the installation, operation, maintenance, training, and support of weapon systems, weapon systems components, and support equipment. Technical manual specifications do not require the preparation of DIDs, but must be approved and assigned an Acquisition Management Systems Control number by the cognizant DID Approval Authority.

Technical Order (Control Manual)

A control manual identifies all depot overhaul and repair tasks recorded in a preferred sequence, support equipment (special tools and test equipment), consumables, a list of support data required to accomplish each task, and provides a means of determining the configuration of support data and equipment.

Technical Publications

Those formal technical orders/manuals developed, as well as commercial, advance, real property installed equipment, and miscellaneous manuals for the installation, operation, maintenance, overhaul, training, and reference of hardware, hardware systems and computer programs; and contractor instructional materials, inspection documentation, and historical-type records that may accompany individual items of equipment.

Technical Reviews

A series of system engineering activities by which the technical progress on a project is assessed relative to its technical or contractual requirements. The reviews are conducted at logical transition points in the development effort to identify and correct problems resulting from the work completed thus far before the problems can disrupt or delay the technical progress. The reviews provide a method for the contractor and Government to determine that the development of a configuration item and its documentation have met contract requirements.

Technical Standard

A standard that controls the medium or process of exchanging data between a sending and a receiving system. Data exchange is defined in terms of presentation formats and transformation of those presentation formats. Technical standards include document image standards; separate graphics, test, and alphanumeric standards; and integrated standards.

Telecommunication

Any transmission, emission, or reception of signs, signals, writings, images, and sounds or intelligence of any nature by wire, radio, visual, or other electromagnetic systems.

Telegraphy

A system of telecommunication that is concerned in any process providing transmission and reproduction at a distance of documentary matter, such as written or printed matter or fixed images, or the reproduction at a distance of any kind of information in such a form. For the purposes of international radio regulations however, unless otherwise specified therein, telegraphy means "a system of telecommunications for the transmission of written matter by the use of a signal code."

Telemetering

The use of telecommunication for automatically indicating or recording measurements at a distance from the measuring instrument.

Telemetry Points

Sampling points within a spacecraft from which onboard conditions, such as voltages, temperature, and so forth, are examined and encoded for transmission to the ground.

Telephony

A system of telecommunication setup for the transmission of speech or other sounds.

Tempering

A heat treatment of steels performed after hardening to partially restore ductility. The steel is heated to a temperature below the critical range followed by any desired rate of cooling.

Tempest
Investigations and studies of compromising emanations. Tempest is sometimes used synonymously with the term *compromising emanations*.

Temporary Configuration Change
An authorization to provide a temporary capability to a CI/CSCI in support of a unique mission requirement or a temporary resolution to a safety, maintenance, or contractual schedule problem.

Temporary Marking
Temporary marking is marking that is intended to remain integral with a component, component part, or piping for a limited time only. Temporary marking may be applied for identification, recording inspection information, or dimensional layout or control.

Tensile Bar (Specimen)
A compression or injection molded specimen of specified dimensions used to determine the tensile properties of a material.

Tensile Strength
The resistance to breaking that metals offer when subjected to a pulling stress.

Terminal Area
 - see LAND.

Terminal Diagram
A diagram relating the functionally depicted internal circuit of an item or device to its terminal physical configuration, and locating the terminals with respect to the outline or orientation markings of the item.

Terminal Pad
 - see LAND.

Testability
(1) A design characteristic that allows the status (operable, inoperable, or degraded) of an item to be determined and the isolation of faults within the item to be performed in a timely manner. (2) A design characteristic that allows the status of a unit or system to be determined in a timely and cost-effective manner.

Testability Analysis
An element in equipment design analysis effort related to developing a diagnostic approach and then implementing that approach.

Test Analysis
The examination of the test results to determine whether the device is in a "go" or "no-go" state or to determine the reasons for or location of a malfunction.

Test and Validation
Physical measurements taken to verify conclusions obtained from mathematical modeling and analysis or taken for the purpose of developing mathematical models.

Test Bench
Equipment specifically designed to provide a suitable work surface for testing a unit in a particular test setup under controlled conditions.

Test Board
A printed board suitable for determining acceptability of the board or of a batch of boards produced with the same process so as to be representative of the production board.

Test Coupon
A portion of a printed board or of a panel containing printed coupons, used to determine the acceptability of such a board.

Test Discrepancy
A test discrepancy is a functional or structural anomaly that occurs during testing and that indicates a possible deviation from specification requirements for the test item. A test discrepancy may be a momentary, not repeatable, or permanent failure to respond in the predicted manner to a specified combination of test environment and functional test stimuli. Test discrepancies may be due to a failure of the test unit or to some other cause, such as the test setup, test instrumentation, supplied power, test procedures, or computer software used.

Test Effectiveness
Measures that include consideration of hardware design, BIT design, test equipment design, and test program set design. Test effectiveness measures include, but are not limited to, fault coverage, fault resolution, fault detection time, fault isolation time, and false alarm rate.

Test Entry
Initial entry and subsequent editing of textual material, typified by messages.

Test Equipment
Electric, electronic, mechanical, hydraulic, or pneumatic equipment (either automatic, manual, or any combination thereof) that is required to perform the monitoring, fault detection, and fault isolation functions.

Test Fixture
An item providing electronic, electrical, pneumatic, etc. and mechanical compatibility between the unit under test and the test equipment.

Test Point
(1) Special points of access to an electrical circuit, used for testing purposes. (2) An electrical contact designed into a circuit specifically for the measurement of internal signals so as to increase the testability of the circuit.

Test Procedure
A document that describes step by step the operations required to test a specific item. a test procedure can be UUT oriented or test equipment oriented.

Test Program Set
The combination of test program, interface device, test program instruction, and supplementary data required to initiate and execute a given test of a unit under test (UUT).

Test Provisions
The capability included in the design for conveniently evaluating the performance of (and locating the faulty item in) a system, subsystem, set, group, unit, assembly, or subassembly.

Test Requirements Document
(1) An item specification that contains the required performance characteristics of a unit under test (UUT) and specifies the test conditions, values (and allowable tolerances) of the stimuli, and associated responses needed to indicate a properly operating UUT. (2) A specification document that contains the required performance characteristics and interfacing

requirements for a unit under test and specifies the tests, conditions, values (and associated tolerances) of the stimuli, and associated responses needed to (a) indicate a properly operating UUT, (b) detect and indicate all faults and out of tolerance conditions, (c) adjust and align the UUT (when applicable), and (d) isolate each fault or out of tolerance condition to the assembly indenture level and degree of ambiguity necessary to effect repair or adjustment of the UUT consistent with the established maintenance concept. (Note that unambiguous fault isolation is usually desired but is not always economically justifiable.)

Test Specification
A document that specifies the required performance characteristics, interface requirements, tests, test conditions, and values (and associated tolerances) of the stimuli and associated responses needed to test a unit under test for a particular test objective.

Test Stand
Equipment specifically designed to provide suitable mountings, connections, and controls for testing electrical, mechanical, or hydraulic equipment as an entire system.

Test Strategy
The arrangement of specific tester types to achieve optimum throughput and diagnostic capability at the least possible cost given the fault spectrum process yield, production rate, and product mix for a particular production environment.

Test Support Software
Computer programs used to prepare, analyze, and maintain test software.

Test Validation
A process in the production of a test program by which the correctness of the program is assured by running it on the ATE together with the unit under test.

Test Verification
The Government performance of a test in accordance with a test procedure to demonstrate that the test setup (and the test procedure) covers the test requirements of the unit under test.

Text Element Identifier (TEI)
A four-character mnemonic symbol identifying a data element. The TEI consists of three alphabetical characters followed by a space. Following the TEI will be the relevant data in the assigned data field.

Text Standard
A technical standard describing the digital exchange of textual data.

Thermal Barrier
The speed at which the heating effect upon a missile moving through the atmosphere imposes a limitation to flight. The effect is due to friction heating and is regarded as a range of velocities above which special cooling methods must be applied rather than as a true ceiling.

Thermal Survey
The measurement of thermal response characteristic at points of interest within an equipment when temperature extremes are applied to the equipment.

Thermogalvanic Corrosion
A type of galvanic corrosion resulting from temperature different points on a metal surface.

Thermoplastic
A plastic that repeatedly can be softened by heating and hardened by cooling through a temperature range characteristic of the plastic and, when in the softened stage, can be shaped by flow into articles by molding or extrusion.

Thick Film
A film deposited by a screen-printing process and fused by firing into its final form.

Thick Film Circuit
A microcircuit in which passive components of a ceramic-metal composition are formed on a suitable substrate by screening and firing.

Third Angle Projection
Third angle projection is the formation of an image or view upon a plane of projection placed between the object and the observer. Third angle projection is the accepted method used in the United States.

Thread
A thread is a portion of a screw thread encompassed by one pitch. On a single-start thread, it is equal to one turn.

Thread Bolt (ISO Term)
A term used to describe any external thread.

Thread Designation
Abbreviations used as designations for various standard thread forms, thread series, and feature designations for use on drawings. Thread series designations are capital letter abbreviations of names used on drawings, in tables, and otherwise to designate various forms of thread and thread series. They commonly consist of combinations of such abbreviations, which include the names and abbreviations that are now in use, together with references to standards in which they occur for various standard threads.

Thread Nut (ISO Term)
A term used to describe any internal thread.

Thread Runout
 - see VANISH THREAD.

Thread Series
Groups of diameter/pitch combinations distinguished from each other by the number of threads per inch applied to specific diameter.

Threads per Inch
The number of thread pitches per inch. It is the reciprocal of the axial pitch value in inches (1/P).

Threshold
(1) The minimum value of a signal that can be detected by the system or sensor under consideration. (2) A value used to denote predetermined levels, such as those pertaining to volume of message storage, i.e., in-transit storage or queue storage, used in a message switching center. (3) The minimum value of the parameter used to activate a device. (4) The minimum value a stimulus may have to create a desired effect.

Threshold Current
In a laser, the driving current corresponding to lasing threshold.

Threshold Frequency
In optoelectronics, the frequency of incident radiant energy below which there is no photoemissive effect.

Threshold Stress
A limiting stress at which stress-corrosion or hydrogen-stress cracking will develop in a metal in a given exposure period.

Through Connection
An electrical connection between conductive patterns on opposite sides of an insulating base, e.g., plated-through hole or clinched jumper wire.

Thru
Informal usage of the word *through*.

Thrust Chamber
In a liquid rocket engine, the assembly consisting of the injector, nozzle, and combustion chamber in which mixing of liquid propellants takes place to form hot gases, which are ejected to produce a propelling force.

Thumbscrew
An externally threaded fastener whose threaded portion is of one nominal diameter. It may have an unthreaded portion with a diameter less than, equal to, *or* greater than the diameter of the threaded portion. It has either a vertically flattened, circular knurled, or wing type head, all of which are designed for rotation by the thumb and fingers. For items having wrenching facilities such as socket recess, multiple spline, or slot heads, use screw (as modified) *or* bolt (as modified).

Thunderstorm Day
A local calendar day during which thunder is heard.

Thyristor
A bistable semiconductor device that comprises three or more junctions and can be switched from the off state or on state to the opposite state.

Time Ambiguity
A situation in which more than one different time or time measurement can be obtained under the stated conditions.

Time Instability
The fluctuation of the time interval error caused by the instability of a real clock.

Time Interval Error
After a time period following perfect synchronization between a real clock and an ideal uniform time scale, the time difference between that clock and the time scale.

Time Scale
(1) A time-measuring system used to relate the time of occurrence of events. (2) Time coordinates placed on the abscissa of Cartesian graphs used to depicting waveforms and similar phenomena.

Tinning
A process for the application of solder coating on component leads, conductors, and terminals to enhance solderability.

Title Block
The block located in the lower right corner of the drawing format, which contains the primary drawing identification.

Toe Crack
A crack in the base metal occurring at the toe of a weld.

Toe of Weld
The junction between the face of a weld and the base metal.

Tolerance
The total amount by which a specific dimension is permitted to vary. The tolerance is the difference between the maximum and minimum limits.

Tolerance, Bilateral
A tolerance in which variation is permitted in both directions from the specified dimension.

Tolerance, Geometric
The general term applied to the category of tolerances used to control form profile, orientation, location and runout.

Tolerance Limit
A tolerance limit is the variation, positive or negative, by which a size is permitted to depart from the design size.

Tolerance, Unilateral
A tolerance in which variation is permitted in one direction from the specified dimension.

Tolerance Zone
The zone between the maximum and minimum limits of size.

Tooling Holes
The general term for the holes placed on a printed board, or a panel, and used to aid in the manufacturing process.

Topside Areas
All shipboard areas continuously exposed to the external electromagnetic environment, such as the main deck and above, catwalks, and those exposed portions of gallery decks.

Total Program Needs
A general term used for the life-cycle requirements and considerations given to the development, production, operations, and support of a configuration item. Total program needs include, but are not limited to, quality assurance; reliability; maintainability; producibility; test and evaluation; acceptance; approval for production; integrated logistic support; personnel; training; availability; interoperability; interchangeability; transportability; survivability; nuclear biological, and chemical hardening; operational readiness; security; safety; schedules; competitive reprocurement; and total life-cycle costs.

Total Thread
Includes the complete and all of the incomplete thread, thus including the vanish thread and the lead thread.

Traceability
The ability to relate individual measurement results to national standards or nationally accepted measurement systems through an unbroken chain of comparisons.

Tracing
A translucent master suitable for reproduction and on which the subject matter is rendered in pencil or ink.

Trade-off
A comparison of two or more ways of doing something in order to make a decision. Decision criteria normally are quantitative.

Transducer
A device that converts the energy of one transmission system into the energy of another transmission system.

Transfer Molding
A method of molding thermosetting materials in which the plastic is first softened by heat and pressure in a transfer chamber, then forced at high pressure through suitable sprues, runners, and gates into a closed mold for final curing.

Transfer Process
Silver halide photo-duplication method involving transfer of the photographic image from the surface of the photosensitive material to a sheet of prepared paper, by contact techniques.

Transient Radiation Effects on Electronics (Tree)
Effects on electronics resulting from a nuclear event. In sensitive semiconductors, the energy absorbed in electronic parts may be sufficient to temporarily after or permanently alter the operating characteristics/state of the semiconductor device.

Transient
A transient is that part of the variation in a variable that ultimately disappears during transition from one steady-state operating condition to another. Transients are aperiodic and include nonrecurring current surges and voltage spikes.

Transient Sounds
Sounds that occur during turn-on and turn-off of the equipment and infrequent sounds that are less than 15 seconds in duration. If sounds occur at intervals of 1/2 second or less, the sound is considered steady-state sound.

Transillumination
Light passed through, rather than reflected off, an element to be viewed, e.g., illumination used on console panels or indicators utilizing edge and/or backlighting techniques on clear, translucent, fluorescent, or sandwich-type plastic materials.

Translation
In geometric dimensioning and tolerancing it is a symbol that indicates that a datum feature simulator is not fixed at its basic location and shall be free to translate.

Transistor
An active semiconductor device capable of providing power amplification and having three or more terminals.

Transition Fit
A transition fit is one having limits of size so prescribed that either a clearance or interference may result when mating parts are assembled.

Transonic
Pertaining to the speed of a vehicle moving through a surrounding fluid, when certain portions of the vehicle are traveling at subsonic speeds while other portions are moving at supersonic speeds.

Transpassive
A noble region of potential where an electrode exhibits a higher than passive current density.

Transponder
Usually, a beacon containing a receiver and a transmitter. The purpose is to increase the energy level of the target or missile signal so that the surveillance range can be extended.

Transportation Control Number (TCN)
The single standard shipment identification number for all DoD-sponsored movements (i.e., materiel and equipment and all vendor shipping transactions involving DoD materiel). The TCN is a 17-position alpha-numeric data element assigned to control a shipment unit through the transportation system (to include CONUS shipments, shipments entering the DTS, and commercial systems).

Tread Plate
Sheet or plate having a raised figured pattern on one surface to provide improved traction.

Triboelectric Effect
The generation of electrostatic charge on an object by rubbing or other type of contact.

Trim Lines
Lines that define the borders of a printed board. See also Corner Mark.

Trimetric Projection
A trimetric projection is an axonometric projection in which all three axes of the object make unequal angles with the plane of projection. The axes make three different angles with each other on the drawing.

True Geometric Counter Part
The theoretically perfect boundary (virtual condition or actual mating envelope) or best-fit (tangent) plane of a specified datum feature.

True Geometry Views
Views that show the actual shape description, and when it is a section view it shows the actual shape cut by the cutting plane.

True Position
The theoretically exact location of a feature established by basic dimensions. Obsolete term; see Positional Tolerance.

Tube
A hollow wrought product that is long in relation to its cross section, which is symmetrical and is round, a regular hexagon or octagon, elliptical, or square or rectangular with sharp or rounded corners, and that has uniform wall thickness except as affected by corner radii.

Tube Bend Drawing
A tube bend drawing establishes, by pictorial or tabular delineation or a combination thereof, end product definition for a single, multiplane, tube, or tube assembly (tube with end fittings). It establishes item identification for the bend tube or tube assembly.

Tube-to-Header "Seal" Welds
Welds between various types of boiler tubes and their respective headers (or drums), such as economizer headers, superheater headers, and so forth. These welds are located on the interior of the header (or drum). Integrity of the tube-to-header connection is provided by a combination of welding and rolling the tube.

Twist
The deformation parallel to a diagonal of a rectangular sheet such that one of the corners is not in the plane containing the other three corners.

Twisted Pair
TWPR is the proper abbreviation.

Type
This term implies differences in like items or processes relative to design, model, shape, etc. and is usually designated by Roman numerals, thus "Type I," "Type II," etc.

Type Designation
A combination of letter and numerals arranged in a specific sequence to provide a short significant method of identification.

U

Undercut (After Fabrication)
The distance on one edge of a conductor measured parallel to the board surface from the outer edge of the conductor, excluding overplating and coatings, to the maximum point of indentation on the same edge.

Undimensioned Drawing
An undimensioned drawing portrays the shape and other design features of an object at a precise scale predominantly without dimensions. It provides an accurate pattern of the feature or features of an item.

Unilateral Tolerance
A tolerance in which variation is permitted in one direction from the specified dimension.

Unique Identification (UID)
A system of establishing globally unique and unambiguous identifiers within the Department of Defense, which serves to distinguish a discrete entity or relationship from other like and unlike entities or relationships.

Unique Item Identifier (UII)
A globally unique and unambiguous identifier that distinguishes an item from all other like and unlike items. The UII is a concatenated value that is derived from a UII data set of one or more data elements. See Part Number or Identifying Number and Item Identification.

Unique Item Identifier (UII) Data Set
A set of one or more data elements marked on an item from which the concatenated UII can be derived. The UII is limited to 50 characters.

Unit
(1) An assembly or any combination of parts, subassemblies, and assemblies mounted together and normally capable of independent operation in a variety of situations (e.g., hydraulic jack, electric motor, electronic power supply, internal combustion engine, electric generator, radio receiver). This term replaces the term *component*. *Note*: the size of an item is a consideration in some cases. An electric motor for a clock may be considered a part, because it is not normally subject to disassembly. (2) A self-contained collection of parts and/or assemblies within one package performing a specific function or group of functions, and removable as a single package from an operating system.

Unit of Issue (UI)
The UI is a standard or basic quantity that is expressed as a unit and indicated in a requisition, contract, or order as the minimum quantity issued (bottle, can, dozen, each, foot, gallon, gross, pair, pound, yard, etc.).

Unit of Issue: Definitive
A definitive UI is a type of UI designation that indicates an exact quantity of volume, linear measurement, weight, or count (e.g., assembly, each, kit, set, foot, etc).

Unit of Issue: Nondefinitive
A nondefinitive UI is a type of UI designation that does not indicate an exact quantity of volume, linear measurement, weight, or count such as drum, can, box, or roll. When a nondefinitive UI is specified, it is accompanied by a quantitative expression (1 RO (150 feet) or 1 RL (50 feet)).

Unit of Product
The unit of product is the thing inspected in order to determine its classification as defective or nondefective or to count the number of defects. It may be a single article, a pair, a set, a length, an area, an operation, a volume, a component of an end product, or the end product itself. The unit of product may or may not be the same as the unit of purchase, supply, production, or shipment.

Unit Pack
The first tie, wrap, or container applied to a single item, or a quantity thereof, or to a group of items of a single stock number, preserved or unpreserved, that constitutes a complete or identifiable package.

Unit Under Test (UUT)
A UUT is any system, set, subsystem, assembly, or subassembly undergoing testing.

Unsupported Hole
A hole containing neither conductive material nor any other type of reinforcement.

Upset Head (Shop Formed)
The head formed during the riveting operation.

U.S.
The abbreviation used on items to denote Government ownership and to comply with public law or other Government regulations. Alternative version is "US," without periods.

Use-As-Is
A disposition of material with one or more minor non-conformances determined to be usable for its intended purpose in its existing condition.

Used with, but Not Part of
A listing of equipment that the item is normally used with, but issued as "part of".

User
A person, organization, or other entity that employs the services provided by a communications system for transfer of information to others.

User-Computer Interface (UCI)
The modes by which the human user and the computer communicate information and by which control is commanded, including areas such as: information presentation, displays, displayed information, formats and data elements; command modes and languages; input devices and techniques; dialog, interaction and transaction modes; timing and pacing of operations; feedback error diagnosis, prompting, queuing and job performance aiding; and decision aiding.

User Interface
The set of hardware, software, and procedures that enable a user to give directions to a computer and receive information from a computer e.g., command language, function key assignments, menus, icons, mouse, and other similar input and output devices.

U.S. Military Property
Government owned property within DOD jurisdiction.

V

Vacuum Bag Molding
A process in which the layup is cured under pressure generated by drawing a vacuum in the space between the layup and a flexible sheet placed over it and sealed at the edges.

Validation
(1) That process in the production of a test program by which the correctness of the program is verified by running it on the automatic test equipment together with the unit under test. The process includes the identification of run-time errors, procedure errors, and other non-compiler errors, not covered by pure software methods. (2) The effort required of the contractor or preparing activity, during which the technical data product is tested for technical adequacy, accuracy, and compliance with the provisions of the specifications and other contractual requirement. Validation is accomplished by comparing the data product with the actual systems or equipment for which the data product was prepared. Validation is normally conducted at the preparing activity or vendor's facility. In extenuating circumstances, validation may be conducted at an alternative site. (3) Confirmation by examination and provisions of objective evidence that the particular requirements for a specific intended use have been fulfilled; that all requirements have been implemented correctly and completely and are traceable to system requirements.

Validation Phase
The second phase in the materiel acquisition process. This phase consists of those steps necessary to resolve or minimize special logistic problems identified during the conceptual phase, verify preliminary design and engineering, accomplish necessary planning, fully analyze trade-off proposals, and prepare contracts as required for full-scale development.

Value Engineering
A disciplined, organized methodology for defining functions, establishing cost/function relationships, and performing trade-off adjustments to realize a system of high value.

Vanish Thread (Partial Thread, Washout Thread, or Thread Runout)
That portion of the incomplete thread that is not fully formed at root or at crest and root. It is produced by the chamfer at the starting end of the thread forming tool.

Variable Item Levels
Variable systems, subsystems, centrals, centers, sets, groups, and units are those configurations whose scope or functions may be varied through the additions or nomenclature must show at least one item of varying quantity or have a variable item therein.

Variable Unit
A unit whose capabilities or functions may be varied through the addition or deletion of assemblies, subassemblies or parts.

Vapor (or Volatile) Corrosion Inhibitor
A chemical that vaporizes and condenses on nearby surfaces, retarding corrosion from moisture.

Variability
The natural tendency for a characteristic of a product, process, or service to differ from a norm or specification target.

Variability Reduction
A planned effort to decrease variability of selected key characteristics.

Variation
The extent to which a product or service is unlike a given standard. This difference can be traced by to sources such as management, product/process specifications, component specifications, poor supplier materials, operator errors, etc.

Vector Graphics
The presentation or storage of images as sequences of line segments.

Vendor
A source from whom a purchased item is obtained; used synonymously with the term *supplier*.

Vendor Developed Item
A specialized version of a vendor's general product line which is not normally stocked as an "off-the-shelf item" but is procurable on order.

Vendor Item Drawing
A vendor item drawing depicts an existing commercial item or vendor-developed item advertised or catalogued as available on an unrestricted basis, on order as an "off-the-shelf " item or an item that, while not commercially available, is acquirable on order from a specialized segment of an industry. A vendor item drawing provides an engineering description and acceptance criteria for purchased items. It provides a list of suggested suppliers, the supplier's item identification, and sufficient engineering definition for acceptance of interchangeable items within specified limits. The vendor item drawing, with any applicable dash numbers, establishes administrative control numbers for identifying the items on engineering documentation. Previously known as *Specification Control Drawing*.

Vendor Substantiation Data
The quality conformance inspections, tests, evaluation criteria, and procedures to be followed to approve a potential supplier as a qualified source for a product, material, or process used in an aerospace propulsion system.

Ventral
Pertaining to the belly or underside of a missile or aircraft, as in *ventral fin*.

Verification
(a) A review process to ensure that a deliverable meets all requirements stipulated in the contract, is in compliance with applicable DOD standards and specifications (unless waived by contract), and is complete and consistent with hardware/product configuration. (b) Confirmation by examination, and provisions of objective evidence, that the item identification marking requirements specified and the associated contract have been fulfilled.

Vernier Scale
A two-piece measuring device generally found on a bellows (large-format) camera to indicate object distance.

Version

(1) An identified and documented body of software. (2) An identified and documented body of software. Modifications to a version of software (resulting in a new version) require configuration management actions by either the contractor, the Government, or both.

Version Description Document (VDD)

The version description document describes the exact release and version of an individual computer software configuration item or a major software development. The VDD identifies the software units and components involved, all engineering change proposals (ECPs) incorporated, problems, and known errors that are corrected by the ECPs. The VDD also provides installation instructions and other data needed to load, operate, or regenerate the delivered software.

Vertex Angle

In an optical fiber, the angle formed by the extreme bound meridional rays accepted by the fiber, or emerging from it, equal to twice the acceptance angle; the angle formed by the largest cone of light accepted by the fiber or emitted from it.

Virtual Condition

(1) A constant boundary generated by the collective effects of a size feature's specified maximum material condition or least material condition. (2) A constant boundary generated by the collective effects of a considered feature of the size's specified MMC or LMC and the geometric tolerance for that material condition.

Very High Frequency (VHF)

Frequencies from 30 to 300 MHz.

Very Low Frequency (VLF)

Frequencies from 3 to 30 kHz.

Via Hole

A plated-through hole used as a through connection, but in which there is no intention to insert a component lead or other reinforcing material.

Via Net Loss (VNL)

Pertaining to circuit performance prediction and description that allows circuit parameters to be predetermined and the circuit to be designed to meet established criteria by analyzing actual, theoretical, and calculated losses.

Vibration Survey

The measurement of vibration response characteristics at points of interest within equipment when vibration excitation is applied to the equipment.

Virtual Condition

The boundary generated by the collective effects of the specified MMC limit of size of a feature and by any applicable geometric tolerances.

Viscosity

That property of a material by virtue of which it tends to resist deformation or flow.

Void

The absence of substances in a localized area.

Voltage

The term most often used in place of electromotive force, potential, or voltage drop, to designate electric pressure that exists between two points and is capable of producing a flow of current when a closed circuit is connected between the two points.

Voltage Drop

The amount of voltage loss from original input in a conductor of given size and length.

Voltage Plane

A conductor portion of a conductor layer on or in a printed board that is maintained at other than ground potential. It can also be used as a common voltage source, for heat sinking, or for shielding.

Voltage Plane Clearance

Voltage plane clearance is the etched portion of a voltage plane around a plated-through or nonplated-through hole that isolates the voltage plane from the hole.

Voltage Standing Wave Ratio (VSWR)

The ratio of maximum to minimum voltage in a standing wave pattern that may appear along a transmission line. *Note*: it is used as a measure of impedance mismatch between the transmission line and it load.

W

Waiver
A written authorization to accept an item, which during manufacture, or after having been submitted for Government inspection or acceptance, is found to depart from specified requirements but nevertheless is considered suitable for use "as is" or after repair by an approved method.

Walk-Through
(1) Formal meeting sessions for the review of source code and design by the various members of the project, for technical rather than management purposes. The purpose is for error detection and not correction. (2) A formal, multidisciplinary paper design review of a computer program, specification, structure, program, logic, modules, code, etc., often using hypothetical inputs.

Wall
A solid feature at any physical orientation composed of opposing surfaces having a nominally uniform thickness.

Wall Thickness
The actual local size between all sets of opposing points on the surface of a wall.

Warning Signal
A signal that alerts the operator to a dangerous condition requiring immediate action.

Warp
- see BOW.

Warranty
The contractual agreement between the Government and the contractor relative to the nature, usefulness, or condition of the item furnished under the contract. Warranty duration is expressed in terms of hours, days, months, number of operations, etc. Warranty markings give notice to a user whether the item is subject to the warranty provisions.

Warranty Markings
Markings that apply when a shipment contains items with a service life defined in a specific amount of hours, specific end date, or a specific operating time.

Washout Thread
- see VANISH THREAD.

Water-Deluge System
A high-capacity, high-pressure system in which water is used at the test and launch stands to cool the missile system and the immediate area during engine operations, including launchings.

Waterproof
(1) Extreme resistance to damage or deterioration by water in liquid form. (2) Exerts a barrier effect to passage of water vapor.

Water-Resistant
Having a degree of resistance of permeability of and damage caused by water in liquid form. Many materials called "waterproof " should correctly be termed "water-resistant to a certain degree."

Water-Tight
That quality of a container or package by which it prevents the passage of liquid water either into or out of the package.

Water Vapor Proof
(1) Not subject to damage by water vapor. (2) Resistant to passage of water vapor, though not necessarily a complete barrier.

Waveguide
A metal pipe of circular or rectangular cross section used for conducting UHF radio waves because of its low loss by attenuation and radiation.

Wave Impedance
The ratio of the electric field strength to the magnetic field strength at the point of observation.

Wavelength
(1) Distance traveled by a periodic disturbance in one period or cycle. (2) Distance between corresponding phases of two consecutive waves of a wave train. (3) Quotient of velocity of propagation divided by frequency. (4) Distance between points having corresponding phase in two consecutive cycles of a periodic wave.

Wave Soldering
A process wherein printed boards are brought in contact with the surface of continuously flowing and circulating solder.

Wave Trap
Device used to exclude unwanted radio signals or interference from a receiver.

Waviness
Waviness is the more widely spaced component of surface texture. Unless otherwise noted, waviness is to include all irregularities whose spacing is greater than the roughness sampling length and less than the waviness sampling length. Waviness may result from such factors as machine work deflections, vibrations, chatter, heat treatment, or warping strains. Roughness may be considered superposed on a "wavy" surface.

Weapon Replaceable Assembly (WRA)
A generic term which includes replaceable packages of a system installed in the weapon system with the exception of cables, mounting provisions, and fuse boxes or circuit breakers.

Wear Allowance
Extra material left on gaging surfaces to accommodate the wear expected during the useful life of a gage.

Wear-Out Failure
A failure that occurs as a result of deteriorating or mechanical wear and whose probability of occurrence increases with time. Wear-out failures generally occur near the end of the life of an item and are usually characterized by chemical or mechanical changes. These failures frequently can be prevented by adopting an appropriate replacement policy based on the known wear-out characteristics of the item.

Weave Exposure
A surface condition of base material in which the unbroken fibers of woven glass cloth are not completely covered by resin.

Wedge
A prism with a very small angle between the refracting surfaces. Wedges may be circular, oblong, or square in outline.

Weight, Gross
The weight of a complete package, ready for shipment, comprising the commodity, inner container, all packaging material, and the outer container.

Weight, Net
The weight of the commodity alone, excluding the weight of all packaging material or containers.

Weight, Tare
The weight of the container or packaging materials. When a container is filled, or partially filled, the weight of the contents is termed the *net weight*. The weight of the container and packaging materials is the *tare*. The net weight plus the tare weight is the *gross weight*.

Weld
A localized coalescence of metal wherein coalescence is produced by heating to suitable temperatures, with or without the application of pressure, and with or without the use of filler metal. The filler metal either has a melting point approximately the same as the base metal or has a melting point below that of the base metals but above 800 degrees F.

Weld Decay
A localized corrosion of weld metal.

Weld Deposited Hard Surfacing
Weld deposited hard surfacing is weld metal that is deposited for the purpose of providing wear resistance.

Weld Deposited Overlay Cladding
Weld deposited overlay cladding is weld metal that is deposited for the purpose of corrosion protection only.

Weld Symbols
Weld symbols are drawing "on" the reference line.

Welded Fabrication or Weldment
Welded fabrication or weldment refers to any assembly where component parts are joined by welding.

Welded Tube
A tube produced by forming and seam welding sheet longitudinally.

Welding Deposited Buttering
Welding deposited buttering is weld metal deposited on base metal prior to the completed weld to permit the final portion of a dissimilar metal weld to be completed as a similar metal weld.

Welding Procedure
A welding procedure is written instructions designed for use in production welding and repair welding, delineating all the essential elements and guidance to produce reliable welds.

Welding Symbols
He welding symbol consists of several elements. Only the reference line and arrow are required elements. Additional elements may be included to convey specific welding information.

Weldment
An assembly whose component parts are joined by welding.

Wet Layup
A method of making a reinforced product by applying a liquid resin system while the reinforcement is put in place.

Wet Strength
The strength of an organic matrix composite when the matrix resin is saturated with absorbed moisture.

Wetting
The formation of a relatively uniform, smooth, unbroken, and adherent film of solder to a base material.

Wet Winding
A method of filament winding in which the fiber reinforcement is coated with the resin system as a liquid just prior to wrapping on a mandrel.

Whip Antenna
A flexible rod antenna, usually between 1/10 and 5/8 wavelength long, supported on a base insulator.

Whisker
A slender acicular (needle-shaped) metallic growth on a printed board.

White Area
The area or population that does not receive interference-free primary service from an authorized AM station or does not receive a signal strength of at least 1 mV/m from an authorized FM station.

White Noise
Noise having a frequency spectrum that is continuous and uniform over a specified frequency band.

Wicking
Capillary absorption of liquid along the fibers of the base material.

Wideband
(1) The property of any communications facility, equipment, channel, or system in which the range of frequencies used for transmission is greater than 0.1% of the midband frequency. *Note: wideband* has many meanings, depending on the application. *Wideband* is often used to distinguish something from *narrowband*, where both terms are subjectively defined relative to the implied context. (2) In communications security systems, a bandwidth exceeding that of a nominal 4-kHz telephone channel. (3) The property of a circuitry that has a bandwidth wider than normal for the type of circuitry, frequency of operation, or type of modulation. (4) In telephony, the property of a circuit that has a bandwidth greater than 4 kHz. (5) Pertaining to a signal that occupies a broad frequency spectrum.

Will
Establishes a declaration of purpose on the part of the design activity.

Windowing
An orientation for display framing in which the user conceives of the display frame as a window moving over a fixed array of data. The opposite of *scrolling*.

Wire
A single metallic conductor of solid, stranded, or tinsel construction, designed to carry current in an electric circuit, but not having a metallic covering, sheath, or shield.

Wire Data List
Tabular listing indicating point-to-point wire runs and connections of an interface adapter, UUT or other device.

Wire Frame
A type of modeling that represents an object by its edges, forming an "outline" of the object incurve segments.

Wire Harness
A wire harness consists of one or more conductors, including coaxial cables, which are grouped together or treated as a separate assembly for the purpose of ease of assembly or installation.

Wire List
A wiring (wire) list consists of tabular data and instructions necessary to establish wiring connections. A wiring list is a form of connection or interconnection diagram. When the wiring list includes material and such material is not called out on the assembly drawing, the wiring list establishes item identification for the wires as a bundle or kit or wires.

Wire Rope
A group of strands helically twisted or laid about a central core is designed as a wire rope. The strands and/or the core act as a unit.

Wire Segment
A length of wire that is continuous and unbroken between its two intended points of termination. A wire segment that has been broken and then repaired is still considered to be one wire segment.

Wire Wrap
- see SOLDERLESS WRAP.

Wiring
Wires, cables, groups, harnesses, and bundles, and their terminations, associated hardware, and support, installed in a vehicle.

Wiring Devices
Wiring devices are the accessory parts and materials that are used in the installation of wiring, such as terminals, connectors, junction boxes, conduit, clamps, insulation, and supports.

Wiring Diagram (Connection Diagram)
A diagram that shows the connections of an installation or its component devices or parts. It may cover internal or external connections, or both, and contains such detail as is needed to make or trace connections that are involved. The Connection Diagram usually shows general physical arrangement of the component devices or parts.

Wiring Harness Drawing
A wiring harness drawing specifies the engineering requirements and establishes item identification for a wiring harness (a group of individually insulated conductors, including shielded wires and coaxial cables, held together by lacing cord or other binding). A wiring harness may or may not terminate in connectors, terminal lugs, or other similar fittings, and may include small electronic parts.

Word
A character string or a bit string considered to be an entity for some purpose. In telegraph communications, six character intervals are defined as a word when computing traffic capacity in words per minute, which is computed by multiplying the data signaling rate in baud by 10 and dividing the resulting product by the number of unit intervals per character.

Word Length
The number of characters or bits in a word.

Working Data
Data for work in progress that has not been "formally" submitted to the Government but may be provided for information purposes with the understanding that it is preliminary and subject to further iteration.

Work Package
A group of related items (which do not make up a complete assembly) with instructions for installing the items in a major assembly structure (e.g., a power supply and mounting hardware with instructions for installing them in a telecommunication satellite structure).

Work Space
That portion of main storage used by a computer program for temporary storage of data.

World Time
Coordinated universal time.

World Wide Web (WWW)
An international, virtual network-based information service composed of Internet host computers that provide online information in a specific hypertext format. *Note 1:* WWW servers provide Hypertext Markup Language (HTML) formatted documents using the Hypertext Transfer Protocol (HTTP). *Note 2:* Information on the WWW is accessed with a hypertext browser such as Explorer, Firefox, and Safari. *Note 3:* No hierarchy exists in the WWW, and the same information may be found by many different approaches.

Woven Fabric Composite
A major form of advanced composites in which the fiber constituent consists of woven fabric. A woven fabric composite normally is a laminate composed of a number of laminae, each of which consists of one layer of fabric embedded in the selected matrix material. Individual fabric laminae are directionally oriented and combined into specific multiaxial laminates for application to specific envelopes of strength and stiffness requirements.

Wrinkle
A wrinkle is a crease or fold in one or more outer layers of a laminated plastic (copper foil, fabric, paper, conformal coating, etc.) that has been pressed in.

Wrought Iron
A ferrous material, aggregated from a solidifying mass of pasty particles of highly refined metallic iron, with which (without subsequent fusion) is incorporated a minutely and uniformly distributed amount of slag.

X

X-axis

The horizontal axis of a cathode-ray oscilloscope, as distinguished from the vertical, or Y-axis, and the intensity modulation, or Z-axis. It is usually an adjustable linear time base. The reference axis in a quartz crystal.

XOFF

An abbreviation for the ASCII transmission-control character meaning "transmitter off."

XON

An abbreviation for the ASCII transmission-control character meaning "transmitter on."

Y

Yagi Antenna
A linear end-fire antenna, consisting of three or more half-wave elements (one driven, one reflector, and one or more directors). *Note 1:* A yagi antenna offers very high directivity and gain. *Note 2:* The formal name for a *Yagi antenna* is *Yagi-Uda array*.

Yarn, Plied
Yarn made by collecting two or more single yarns together. Normally, the yarns are twisted together, although sometimes they are collected without twist.

Y-axis
The vertical axis on a cathode-ray oscilloscope. A line perpendicular to two opposite parallel faces of a quartz crystal.

Yield Point
The first stress in a material at which an increase in strain occurs without an increase in stress. (The stress is less than the maximum attainable.) It should be noted that only materials that exhibit the unique phenomenon of yielding have a "yield point."

Yield Strength
The stress at which a material exhibits a specified limiting deviation from the proportionality of stress to strain.

R. Hanifan, *Concise Dictionary of Engineering: A Guide to the Language of Engineering,* 265
DOI 10.1007/978-3-319-07839-7_25, © Springer International Publishing Switzerland 2014

Z

Z-State
Or high impedance state. A logic level that a node is set at when all outputs connected to that nodes are deactivated or turned off.

ZEPP Antenna
A horizontal antenna that is a multiple of a half wavelength long and is fed at one end by one lead of a two-wire transmission line that is some multiple of a quarter wavelength long.

Zero Beat
The conditions wherein two frequencies that are being mixed are exactly the same and therefore produce no beat note.

Zero Bias
In a vacuum tube, a condition in which there is no potential difference between the control grid and the cathode.

Zero Compression
Any of a number of techniques used to eliminate the storage of nonsignificant leading zeros during data processing in a computer.

Zero Delay Simulation
A digital logic simulation technique which assumes that all circuit devices have no propagation delay.

Zero Launch
The launch of a missile from a launching platform that has zero-length guide rails.

Zero Length Launcher
A launcher that holds a missile or rocket vehicle in position and releases the vehicle simultaneously at two points so that the buildup of thrust normally rocket motor thrust, is sufficient to take the missile or vehicle off directly into the air without need of a takeoff run and without imposing pitch rate release.

Zeus Trojan
Also known as Zbot, is a malware tool-kit that allows a cybercriminal to build his own Trojan Horse. On the Internet, a Trojan Horse is programming that appears to be legitimate but actually hides an attack.

Zip Drive
A Zip drive is a small, portable disk drive used primarily for backing up and archiving personal computer files. Zip drives and disks come in two sizes. The 100 megabyte size actually holds 100,431,872 bytes of data or the equivalent of 70 floppy diskettes. There is also a 250 megabyte drive and disk.

Zombie
A zombie (also known as a bot) is a computer that a remote attacker has accessed and set up to forward transmissions (including spam and viruses) to other computers on the Internet. The purpose is usually either financial gain or malice. Attackers typically exploit multiple computers to create a botnet is also known as a zombie army.

R. Hanifan, *Concise Dictionary of Engineering: A Guide to the Language of Engineering*, DOI 10.1007/978-3-319-07839-7_26, © Springer International Publishing Switzerland 2014

Zombie Army
A botnet (also known as a zombie army) is a number of Internet computers that, although their owners are unaware of it, have been set up to forward transmissions (including spam or viruses) to other computers on the Internet. Any such computer is referred to as a zombie—in effect, a computer "robot" or "bot" that serves the wishes of some master spam or virus originator. Most computers compromised in this way are home-based.

Zone Blanking
Method of turning off the cathode-ray tube during part of the sweep of an antenna.

Zulu Time (Z)
Coordinated universal time. Formerly a synonym for *Greenwich Mean Time*.